Reducing Climate Impacts
in the Transportation Sector

T0185315

Reducing Climate Impacts in the Transportation Sector

Daniel Sperling

Editor

Institute of Transportation Studies, University of California,
Davis, CA, USA

James S. Cannon

Editor

Energy Futures, Boulder, CO, USA

 Springer

Editors

Daniel Sperling
University of California,
Davis Institute of Transportation Studies
Davis CA 95616
USA
dsperling@ucdavis.edu

James S. Cannon
Energy Futures
1460 Riverside Ave.
Boulder CO 80304
USA
jscannon@energy-futures.com

ISBN: 978-1-4020-6978-9 (PB)   e-ISBN: 978-1-4020-6979-6

DOI 10.1007/978-1-4020-6979-6

Library of Congress Control Number: 2008931157

Printed on acid-free paper

9 8 7 6 5 4 3 2 1

springer.com

Contents

Preface and Acknowledgements

Climate change has fully entered the public consciousness. Newspapers barrage readers with stories of shrinking glaciers, disappearing species, and cataclysmic weather. A documentary on climate change wins an Oscar, a Noble Peace Prize is awarded to scientists studying climate change, and arcane scientific debates become front page news. The reality of climate change and the imperative to do something is now widely accepted. But that is where the agreement largely ends. What to do and how fast to do it remains intensely controversial.

Those questions about what to do about transportation to bring it in line with climate goals was the focus of a high level meeting in California in August 2007. Two hundred leaders and experts were assembled from the automotive and energy industries, start-up technology companies, public interest groups, academia, U.S. energy laboratories, and governments from around the world. Three broad strategies for reducing greenhouse gas emissions were investigated: reducing vehicle travel, improving vehicle efficiency, and reducing the carbon content of fuels. This book is an outgrowth of that conference.

The conference was not a one-off event. It was the latest in a series of conferences held roughly every two years on some aspect of transportation and energy policy, always at the same Asilomar Conference Center near Monterey on the California coast. The first conference in 1988 addressed alternative transportation fuels, the last two have focused on climate change. The full list appears below:

I. Alternative Transportation Fuels in the '90s and Beyond (July 1988)
II. Roads to Alternative Fuels (July 1990)
III. Global Climate Change (August 1991)
IV. Strategies for a Sustainable Transportation System (August 1993)
V. Is Technology Enough? Sustainable Transportation-Energy Strategies (July 1995)
VI. Policies for Fostering Sustainable Transportation Technologies (August 1997)
VII. Transportation Energy and Environmental Policies into the 21st Century (August 1999)

The chapters of this book evolved from presentations and discussions at the 11th Biennial Conference on Transportation and Energy Policy.

The conference was hosted and organized by the Institute of Transportation Studies at the University of California, Davis (ITS-Davis) under the auspices of the United States (U.S.) National Research Council's Transportation Research Board—in particular, the standing committees on Energy, Alternative Fuels, and Transportation and Sustainability.

The conference would not have been possible without the generous support of the following organizations: William and Flora Hewlett Foundation, Surdna Foundation, Energy Foundation, Neil C. Otto, U.S. Department of Energy, U.S. Environmental Protection Agency Office of Transportation and Air Quality, U.S. Department of Transportation Center for Climate Change and Environmental Forecasting, Natural Resources Canada, California Department of Transportation, California Energy Commission, California Air Resources Board, and the University of California Davis Sustainable Transportation Center.

The editors also want to acknowledge the Corporate Affiliate Members of ITS-Davis that provide valuable support that allows the ITS the flexibility to initiative new activities and events such as the conference upon which this book is based. Those companies are Nissan, Toyota, Shell, ExxonMobil, Subaru, Pacific Gas & Electric, Mitsui PowerSystems, Chevron, Aramco Services Company, and Nippon Oil Corporation.

The conference program was directed by Daniel Sperling, along with David Burwell, John DeCicco, Carmen Difiglio, Robert Dixon, Duncan Eggar, Lew Fulton, John German, David Greene, Cornie Huizenga, Roland Hwang, Jack Johnston, Robert Larson, Alan Lloyd, Marianne Mintz, Peter Reilly-Roe, Jonathan Rubin, Mike Savonis, Lee Schipper, Christine Sloane, and Steve Winkelman. This committee worked closely in crafting a set of speakers and topics that was engaging and insightful.

Most of all, we want to acknowledge the many attendees of the conference listed in Appendix B. These invited leaders and experts, coming from many parts of the world and many segments of society, enriched the conference with their deep insights and rich experiences.

California, USA   Daniel Sperling
Colorado, USA   James S. Cannon

Contributors

Anup Bandivadekar
Sloan Automotive Laboratory, 31-168, Massachusetts Institute
of Technology, 77 Massachusetts Avenue, Cambridge,
MA 02139-4307, USA

Rex Burkholder
Metro, 600 NE Grand Avenue, Portland, OR 97232, USA

David G. Burwell
BBG Group, 7008 Rainswood ct., Bethesda, Md. 20817, USA

James S. Cannon
Energy Futures, Inc., POB 4367, Boulder, CO 80306, USA

Lynette Cheah
Sloan Automotive Laboratory, 31-153, Massachusetts Institute
of Technology, 77 Massachusetts Avenue, Cambridge,
MA 02139-4307, USA

Gustavo Collantes
John F. Kennedy School of Government, Belfer Center
for Science and International Affairs, 79 John F. Kennedy Street,
Cambridge, MA 02138, USA

Philippe Crist
International Transport Forum, OECD, 2 rue André Pascal,
75775 Paris Cedex 16, France

Mark A. Delucchi
Institute of Transportation Studies, University of California,
Davis (UCD), 2028 Academic Surge, One Shields Avenue,
Davis CA 95616, USA

Christopher Evans
Technology and Policy Program, Massachusetts
Institute of Technology, 77 Massachusetts Avenue, Cambridge,
MA 02139-4307, USA

Carolyn Fischer
Resources for the Future, 1616 P St. NW, Washington, DC 20036, USA

Kelly Sims Gallagher
John F. Kennedy School of Government, Belfer Center for Science and
International Affairs, 79 John F. Kennedy Street, Cambridge, MA 02138, USA

John German
American Honda Motor Company, 3947 Research Park Drive,
Ann Arbor MI 48108, USA

David L. Greene
National Transportation Research Center, ORNL,
2360 Cherakala Boulevard, Knoxville TN 37932, USA

Anthony Greszler
Volvo Powertrain Corp., 10 Darthmouth Drive, Hagerstown MD 21742, USA

John Heywood
MIT/CTPID, E40-227, 77 Massachusetts Avenue, Cambridge MA 02139, USA

Nic Lutsey
Institute of Transportation Studies, University of California,
Davis (UCD), 2028 Academic Surge, One Shields Avenue,
Davis CA 95616, USA

Amy Myers Jaffe
Rice University, MS 40, 6100 Main Street, Houston TX 77005, USA

Eliot Rose
Metro, 600 NE Grand Avenue, Portland, OR 97232, USA

Jack Short
International Transport Forum, OECD, 2 rue André Pascal,
75775 Paris Cedex 16, France

Dan Sperling
Institute of Transportation Studies, University of California, Davis (UCD),
2028 Academic Surge, One Shields Avenue, Davis CA 95616, USA

Kurt Van Dender
International Transport Forum, OECD, 2 rue André Pascal,
75775 Paris Cedex 16, France

Chapter 1
Climate Change and Transportation

Dan Sperling, James Cannon and Nic Lutsey

More than 200 experts and leaders from around the world gathered in August 2007 at the 11th Biennial Conference on Transportation and Energy Policy at the Asilomar conference center in Pacific Grove, California. During three days, they tackled what many agree is the greatest energy and environmental challenge the world faces: climate change. The conference came at a time when the latest report by the United Nations Intergovernmental Panel on Climate Change, the most complete and authoritative scientific assessment to date, raised the spectre of even more dramatic climate changes than had been assumed in the past (IPCC, 2007a). The IPCC, together with former Vice President Al Gore, who starred in an academy award winning documentary, *The Inconvenient Truth*, received the 2007 Nobel Prize for their efforts in highlighting the dangers and risks of climate change.

Most environmental scientists now acknowledge that climate change is a real global problem, and that transportation is a key contributor. But the world is still near the starting line in doing much about it.

The 2007 Asilomar Conference examined the role of transportation in reducing greenhouse gas (GHG) emissions. This book is based on presentations and discussions that took place there. It draws upon the knowledge and insights of the world's experts.

Transportation and Climate Change

Transportation accounts for about one-fifth of global GHG emissions causing climate change, but close to 30 percent in most industrialized countries. The United States far exceeds the rest of the rest of the world when it comes to transport-related GHG emissions. While China, India, and other countries in the developing world are rapidly motorizing, causing rapid increases in their

D. Sperling
Institute of Transportation Studies, University of California, Davis (UCD), 2028
Academic Surge, One Shields Avenue, Davis CA 95616, USA

D. Sperling, J.S. Cannon (eds.), *Reducing Climate Impacts
in the Transportation Sector*, DOI: 10.1007/978-1-4020-6979-6_1,
© Springer Science+Business Media B.V. 2009

Table 1.1 U.S. GHG emissions by energy sector since 1990 (Tg CO2e), (EPA, 2006)

Energy sector	1990	1995	2000	2004	2005
Tranportation	1,467.0	1,593.3	1,787.8	1,868.9	1,897.9
Industrial	1,539.8	1,595.8	1,660.1	1,615.2	1575.2
Residential	929.9	995.4	1,131.5	1,175.9	1,208.7
Comercial	759.2	810.6	969.3	999.1	1,016.8
Total U.S. Territories	28.3	35.0	36.2	54.0	52.5
Total	4,724.0	5,030.0	5,584.9	5,713.0	5,751.2
Total Electrical Generation	1,810.2	1,939.3	2,283.5	2,315.8	2,381.2

GHG emissions, their transport emissions are still a fraction of those in the U.S.

As indicated in Table 1.1, transportation activities accounted for 33 percent of GHG emissions in the United States in 2005. Virtually all of the transportation energy consumed came from petroleum products. Over 60 percent of the emissions resulted from gasoline consumption for personal vehicle use.

As shown in Fig. 1.1, U.S. GHG emissions have varied widely since 1990, but generally have increased about 1 percent per year, roughly half that increase coming from transportation (EPA, 2006).

Transport-related energy use and GHG emissions are expected to continue increasing into the foreseeable future. The U.S. government, in its "Annual Energy Outlook 2008" report, forecasts a 25 percent increase in total oil use between 2006 and 2030, from 20.7 to 24.9 million barrels per day. About 2/3 of that oil will be used for transportation (EIA, 2007).

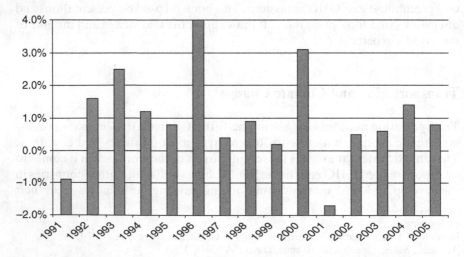

Fig. 1.1 Annual percent change in U.S. GHG emissions since 1990 (EPA, 2006)

Trends in Climate Change

The climate change debate intensified dramatically in 2007. A study by the United Nations Environment Program (UNEP, 2007), prepared by 390 experts and reviewed by more than 1,000 others, concluded that climate change is one of several pressing global problems that are putting the human race at risk. The report warned that unmitigated climate change would, in the long term, likely swamp the capacity of natural, managed and human systems to adapt.

Perhaps most instrumental was the release of the latest IPCC report (IPCC, 2007a). The most complete and authoritative scientific assessment to date, reflecting the views of thousands of climate scientists, it clearly affirmed the role of human activities, primarily fossil fuel burning, in creating climate change. It documented rising air and ocean temperatures, accelerated melting of glacial snow and ice, and slow but steady rising of ocean levels. Eleven of the last 12 years evaluated by the IPCC ranked among the warmest years since 1850.

The evidence of change is powerful and compelling. The IPCC found that average Northern Hemisphere temperatures were higher during the second half of the 20th Century than during any other 50-year period in the last 500 years and likely the highest in at least the past 1,300 years. If global temperatures increase another 1.5–2.5°C, which is likely in the 21st Century, unless GHG emissions are dramatically curtailed, as many as 20–30 percent of plant and animal species are likely to be at increased risk of extinction, according to the IPCC.

Temperature increases will not be uniform. Average temperatures in the Arctic are rising twice as rapidly as in the rest of the world. Satellite data since 1978 show that Arctic sea ice has shrunk in surface area by 2.7 percent per decade, with much greater shrinkage in summer. Equally disconcerting is the rise in sea levels. The global average sea level has risen at an average rate of 1.8 millimeters per year since 1961. Since 1993, the rise has accelerated to 3.1 millimeters per year, caused by melting glaciers, ice caps and polar ice sheets.

The Political Will to Counter Climate Change

There is now little expectation that adaptation or mitigation alone can avoid all climate change impacts. However, they can complement each other and together can significantly reduce the risks of climate change. Adaptation is necessary to address impacts resulting from warming, while early mitigation actions would avoid further locking-in carbon intensive infrastructure and would reduce climate change and associated adaptation needs.

A global attempt to develop and implement a politically viable short term mitigation strategy has been underway for several decades, but with little success. Voluntary reductions in GHG emissions were endorsed by delegates from 189 countries at an international conference held in 1992 in Rio de Janeiro, Brazil.

More serious and mandatory emission reduction targets were incorporated into the 1997 Kyoto Protocol, endorsed by delegates from more than 160 countries meeting in Japan. The protocol took effect in 2005, when countries representing the required 55 percent of global GHG emissions formally signed on.

The global response to climate change set into motion by the Kyoto Protocol has been widespread, even in the United States (Lutsey and Sperling, 2008). But opposition by the United States has continued and rapid growth in GHG emissions in developing nations, particularly in China and India, which are exempted from the protocol's GHG reduction targets, has undermined its effectiveness. By late 2007, it was apparent that the world was not only falling short of complying with the Kyoto Protocol target of a 6–8 percent reduction in GHG gases from 1990 levels, but it was, in fact, still moving in the wrong direction. Global GHG emissions had grown by 20 percent since the Kyoto Protocol's adoption a decade earlier.

Against this backdrop, delegates from 190 countries reconvened at another United Nations Climate Change Conference in Bali, Indonesia in December 2007 to develop a new roadmap. After intense debate that ran a day past the scheduled close of the conference, the group failed to reach a consensus about how best to move ahead after the provisions of the Kyoto Protocol expire in 2012. Instead, the group voted to undertake a set of negotiations aimed at crafting a new international agreement, scheduled to be drafted by 2009.

Combating Climate Changes in the Transportation Sector

GHG mitigation strategies for transportation can be grouped into three categories: vehicle efficiency, low-carbon fuels, and travel reduction. Potential GHG reductions are very large, with varying levels of cost effectiveness. Virtually all provide large co-benefits, including energy cost savings, oil security, and pollution reduction. Table 1.2 categorizes these GHG mitigation options into near and mid-term options and lists key supporting policies and practices needed for their implementation.

Vehicle Efficiency

Available and emerging vehicle efficiency improvements can be categorized into three groups: incremental vehicle technologies, advanced technologies, and on-road operational practices. Incremental improvements include more efficient combustion through such technologies as variable valve systems, gasoline direct injection, and cylinder deactivation; more efficient transmissions, including 5- and 6-speed automatic, automated manual, and continuously variable configurations; use of lightweight materials; and more aerodynamic designs. GHG

Table 1.2 Summary of transportation GHG mitigation options (Lutsey, 2008)

Category	Today's measures (deployable 2007–2015)	Tomorrow's measures (deployable 2010–2030)	Supporting policies and practices
Vehicle efficiency	Incremental efficiency improvements in conventional gasoline automobiles and diesel trucks "On-road" improvements in maintenance practices, technology, driver education and awareness	Increased vehicle electrification (hybrid gas-electric, plug-in hybrid, battery electric) Fuel cell vehicles	Vehicle efficiency performance standards (fuel economy, CO2 emission rate) Voluntary industry commitments Vehicle purchasing incentives (rebates, feebates for low-CO2, high fuel economy) Government and company fleet efficient vehicle purchasing
Low greenhouse gas fuels	Mixing of biofuels in petroleum fuels Use of lower GHG-content fossil fuels (e.g. diesel, compressed natural gas)	Electricity (in plug-in hybrids and battery electrics) Cellulosic ethanol Hydrogen from renewable sources Mobile air-conditioning (MAC) refrigerant replacement	Biofuel blending mandates Low GHG fuel standards Carbon tax on fuels Government and company fleet incorporation of alternative fuels
Vehicle demand reduction	Intelligent transportation system (ITS) technologies to improve system efficiencies Mobility management technologies Inclusion of GHG impacts in land use and transport planning Incentives and rules to reduce vehicle use	Greenhouse gas budgets for households and localities Modal shifts (road to rail freight, public transit systems) ITS technologies to create new more –efficient transport modes	Road, parking, congestion pricing Investment in public transit Public awareness, outreach, education campaigns

emissions rates can be reduced by as much as 30 percent with these approaches. Most studies show that fuel savings more than outweigh the increased vehicle cost when considered over the life of a vehicle (using appropriate discount factors). Similar GHG reductions are possible with commercial freight trucks, also with net cost savings over the life of the vehicle.

Much greater GHG reductions are possible with electric drive propulsion technologies. These include gasoline-fueled hybrid electric vehicles, plug-in hybrids, which use both electricity stored from the grid and petroleum fuels, battery electric vehicles, and hydrogen-powered fuel cell vehicles. These technologies can double vehicle fuel efficiency. When low-carbon electricity, hydrogen and biofuels are used with these vehicles, the lifecycle GHG emissions can be reduced 80 percent or more. However, these advanced technologies involve either larger initial costs for electricity and hydrogen storage or have high development and commercial deployment costs. Because vehicle turnover is slow, it would take a long time to realize these potential reductions.

The third category, on-road efficiency improvements, involves a combination of consumer education, vehicle maintenance practices, and "off-cycle" vehicle technologies. These on-road vehicle efficiency improvements can reduce GHG emissions by up to 20 percent. Improved vehicle maintenance practices for tires, wheels, oil, and air filters can improve vehicle operating efficiencies. Inexpensive new technologies can be added to vehicles to raise driver awareness of fuel use. These include dashboard instruments that display instantaneous fuel consumption, efficient engine operating ranges, shift indicator lights, and tire inflation pressure. Other changes include replacing the conventional air conditioning refrigerant, hydrofluorocarbon HFC-134a, with gases that pose less of a threat to the climate.

A variety of policies aimed at vehicle makers and policies could accelerate these efficiency improvements. These include requirements for more efficient vehicles aimed at automakers and incentives targeted at manufacturers to sell those more efficient vehicles and to consumers to purchase them. If these vehicle policies are linked with actions that increase the supply of low-carbon alternative fuels, as discussed below, the GHG and oil benefits would be still greater.

Low-Carbon Fuels

Increased use of fuels with lower lifecycle GHGs emissions can greatly reduce overall transportation GHG emissions. Most low-carbon transportation fuels face a combination of infrastructural and economic barriers. There are three sets of transportation fuels that have the potential to replace large amounts of petroleum and eliminate large quantities of GHGs. They are biofuels, electricity, and hydrogen.

Biofuels are the easiest since fuels made from food products have been well known for millennia and small amounts can be readily blended into gasoline and diesel fuel. Indeed the United States and many other countries have been doing so for many years, mostly with ethanol made from corn and sugar, but also biodiesel oils extracted from plants and animal fats. Brazil has gone furthest, first using ethanol made from sugar cane in dedicated vehicles in the 1980s and more recently in fuel-flexible vehicles. In Europe, Brazil, and the

United States, biodiesel is used in limited amounts in diesel cars, buses, and trucks. Biodiesel and ethanol, as currently produced, are expensive and divert farmland to energy use, pushing up food prices.

The GHG benefits of ethanol made from sugar cane are substantial, compared to gasoline, but that is not true for ethanol made from corn. In the case of corn, GHGs are reduced only about 10–20 percent, and perhaps not at all if new scientific findings about GHG releases from soils prove correct (Searchinger et al., 2008). Future biofuels, made from cellulosic materials such as grasses and trees, would have much higher lifecycle GHG benefits, especially those made from crop residues and other waste materials. For both GHG and food production reasons, it is entirely possible that the biofuels industry of the future will be based almost solely on waste materials, limiting the scale of potential biofuels production.

Large GHG benefits are possible from hydrogen fuel cells and battery electricity vehicles, including plug-in hybrid electric vehicles that use both gasoline and electricity. If the fuels are obtained from low-carbon feedstocks, such as biomass, wind, or nuclear, or from fossil energy coupled with carbon capture and storage, the result could be tremendous GHG reductions. Both electricity and hydrogen face many barriers, though. All electric vehicle technologies must overcome the high cost and low energy density of batteries, and hydrogen fuel cells must overcome the challenge of jointly deploying an entirely new propulsion technology and fuel.

Alternative fuels have been subsidized and mandated by various governments at various times. A biofuel mandate exists in Europe and ethanol subsidies and mandates have been in place in the United States and Brazil for decades. In December 2007, the United States passed a law requiring 36 billion gallons of biofuels by 2022, including 21 billion gallons of advanced biofuels, expected to be mostly made from cellulosic materials.

A new policy instrument gaining much attention worldwide is the low carbon fuel standard (Farrell and Sperling, 2007). In this case, the government sets a GHG intensity target, for example 10 percent reduction by 2020, and allows companies to meet the requirement however best suits them. Companies are allowed to buy credits when they fall short of the targets and to sell them when they exceed the targets. This innovative approach provides a durable policy framework that can be tightened over time, and avoids the pitfalls of governments picking winners or losers. California adopted this rule in 2007, and many others, including the European Union, are in the process of adopting it as this book goes to press.

The transition to low-carbon alternatives will not be straightforward or unchallenged. Already, the oil industry is investing many tens of billions of dollars in high-carbon unconventional fossil alternatives. These alternatives include tar sands in Canada, very heavy oil in Venezuela, U.S. oil shale, and coal in a variety of countries, especially China, South Africa, and the United States. Fuels made from these sources require much more energy for extraction and processing and therefore have considerably higher GHG emissions than gasoline and diesel fuel made from conventional oil. Only if the carbon is captured at the site and sequestered underground could GHG emissions from these sources be reduced relative to conventional gasoline and diesel fuels.

Travel Reduction

The same technologies and practices implemented by local governments to manage vehicle travel and traffic congestion can also be used to reduce GHG emissions. Strategies to reduce vehicle travel can be sorted into three broad groups: information and communication technologies to provide new and more efficient mobility services; incentives and pricing schemes to encourage less-GHG-intense travel; and denser land use that more efficiently organizes businesses, residences, and services so as to reduce vehicle travel.

Information and communication technologies can be used to simultaneously improve mobility and reduce transport GHG emissions. Incremental enhancements include automating urban traffic signals to streamline traffic and reduce stop-and-go conditions; implementing integrated "smart cards" to facilitate multi-modal travel and increase transit use; and providing real-time traffic data to traffic managers and vehicle users to improve efficiency. More transformational changes are possible that could result in far greater reductions in vehicle travel. These include creating entirely new modes of travel, such as carsharing, paratransit that provides door-to-door service without advanced reservations, and organized ridesharing.

Various incentive and pricing schemes can be designed to reduce GHG-intense travel. Road pricing to reduce congestion in city centers and on clogged highways can smooth flows, encourage transit modes, and reduce vehicle travel. Parking policies that encourage higher occupancy travel modes and internalize the full cost of parking can be highly effective at reducing use of single-occupant vehicles. Workplace incentives to promote telecommuting and carpooling can also help mitigate peak-time congestion travel.

The real key to reduced vehicle travel is creating more choice for travellers, beyond the dominant single-occupant vehicle, and to pursue multiple strategies, especially increased densification of land use. Research shows that residents in more densely populated areas and in areas with better mixes of land uses tend to emit far less GHG emissions from their travel (Boarnet and Crane, 2001; Handy et al., 2007). They tend to walk more, use more public transportation, and drive less. Policies aimed at increasing density and influencing local governments to make land use development and zoning decisions based on likely impact on GHG emissions could be highly effective at reducing emissions. Combined with targeted vehicle and road pricing initiatives, more high quality travel choices, and improved conventional transit services, the result could be a substantial reduction in vehicle travel. At the Asilomar Conference, John Horsely, head of the conservative American Association of Safety and Highway

Officials, announced that his organization now advocates cutting in half the projected increases in vehicle travel. Many believe much larger reductions are possible and desirable (Reid Ewing et al., 2007).

Greenhouse Gas Mitigation Supply Curves

Studies of cost effectiveness generally find transportation GHG reductions more expensive than reductions in most other sectors (IPCC, 2007b; McKinsey, 2007). The high estimated cost is due to low fuel price elasticity by owners of passenger cars and light trucks; strong demand for personal travel; the difficulty of introducing new low-carbon fuels and new fuel-efficient propulsion technologies; deteriorating quality of public transport; and the increasing share of goods carried by truck. In addition, petroleum fuel use is becoming more carbon intense, as easily accessed and high quality reserves are depleted, and as remote sources of unconventional fossil energy are tapped and as additional refining is required to upgrade fuel quality.

On the other hand, many transportation strategies to reduce GHG emissions are highly cost effective. Many generate cost savings over the life of an investment, when future energy savings are calculated using normal discount factors. When other co-benefits are included, such as improved energy security and traffic congestion, many transport GHG mitigation options become attractive. These findings are counter to the conventional thinking that often ignores co-benefits and emphasizes near-term resistance to expanded technology and behavioral options.

GHG mitigation strategies can be ranked using a supply curve framework. They are ranked according to their GHG reduction cost effectiveness, or cost-per-tonne CO_2 equivalent emission reduction. Both the initial costs of the GHG technologies and the lifetime energy savings are included in the cost-per-tonne metric. Co-benefits are usually ignored, but could be added.

Figure 1.2 shows a supply curve of GHG mitigation actions for all sectors of the U.S. economy, with transportation-specific measures highlighted (Lutsey, 2008). The non-transportation actions include electric power sector actions, such as greater use of natural gas, nuclear, and renewable energy, and constructing and retrofitting buildings to be more energy efficient. Other analyses, for instance by McKinsey & Co (2007), find similar relationships.

Whether GHG mitigation is easier or harder in transport than other sectors is an important debate that will continue into the future. What is certain, though, is that many attractive strategies and actions are available. The question is how aggressively the GHG reductions will be pursued by government, industry, and consumers. The existence of large co-benefits, including energy security and oil import reductions, will undoubtedly be influential. There will be many other forces at work, however. Consumers are already altering their behavior to be more environmentally conscious, for instance buying

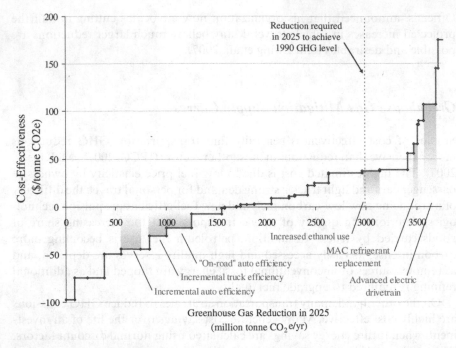

Fig. 1.2 Cost-effectiveness supply curve of available GHG mitigation technologies (Lutsey, 2008)

high-priced hybrid electric vehicles. Moreover, innovations with vehicles, fuels, and new mobility services will undoubtedly lead to new investments in low-GHG options. Competitive forces are at work. Toyota's experience with the Prius hybrid electric car vividly demonstrates the "halo" benefits of being a leader in environmental action. The "halo" created by the successful Prius has increased the attractiveness of Toyota's other vehicles. Companies are increasing their investment in a wide variety of new low-carbon fuels and efficient advanced propulsion technologies to achieve the same halo benefits.

Overview of the Remaining Chapters

Discussions at Asilomar largely followed the topics addressed above, with debates centering on the attractiveness of different policy instruments and differences across the United States., Europe, and the rest of the world. The 10 chapters that follow offer in-depth analyses of many of the most salient issues discussed at Asilomar and increasingly in global public debates. They are authored by presenters or participants at the Asilomar conference, in some cases assisted by colleagues.

The next two chapters set the stage for the discussion of strategies to reduce global climate change from the transportation sector. Amy Myers Jaffe, Associate Director of the Baker Institute at Rice University, examines in her chapter the major supply risks that face international oil markets and considers the carbon emission implications of the kinds of energy supplies that the United States may turn to in an effort to diversify away from rising dependence on Middle East oil. Most of the "easy" oil has already been found, she says, and new supplies are likely to be more difficult to extract technically, be found in more politically problematic countries, and produce more, not less, GHG emissions when processed and burned. Her calculations suggest U.S. energy independence is impossible given the projected growth in domestic oil demand. Therefore, a more ambitious national strategy is needed to address reductions in GHG emissions from transportation. This strategy should address the key international geopolitical issues that undermine energy security as well as climate change.

Jack Short, Secretary General of the International Transport Forum in Paris, France, and his colleagues Kurt Van Dender and Philippe Crist note that actions to combat climate change are now at unprecedented levels. Even so, the growth in GHG emissions from the transportation sector is quickly getting worse. The transportation sector is different in nature and degree from other energy sectors, and it presents unique challenges to policy developers. Their chapter examines present policy measures to reduce CO_2 emissions from private cars in Europe and discusses the implications of tough CO_2 targets for transport policy and for the structure of the transport sector.

The next group of three chapters examine the policies and technologies that are now commercially available or near to commercial viability and that could reduce GHG emissions from motor vehicles. John Heywood at the MIT Laboratory for Energy and Environment and his collaborators, Lynette Cheah, Christopher Evans, and Anup Bandivadekar, examine the vehicle design and sales mix changes necessary to double the average fuel economy or halve the fuel consumption of new light duty vehicles by model year 2035. The analysis concludes that available automotive technologies can do the job, although significant changes in vehicle design are required. There are trade-offs between the performance, cost, and fuel consumption reduction benefits. For example, the extra cost of the 2035 model year vehicles is estimated to be between $54 and $63 billion, or about 20 percent more than the baseline cost. This corresponds to a cost of $65 to $76 per ton of equivalent CO_2 emissions. Heywood et al. warn that the changes required to meet this goal run counter to the trend towards larger, heavier, more powerful vehicles over the last 25 years. Instead, their scenarios depict a transportation future where automakers face higher costs to produce smaller vehicles with performance similar to today's.

John German, Manager of Environmental and Energy Analysis at American Honda Motor Company, discusses technology development for cars and light trucks that meet the needs of customers and the global need to address climate change. He believes that the automotive industry is in a period of unprecedented technology development that will move a long way towards sustainable mobility. Gasoline engine technology is maturing rapidly and manufacturers are working hard on diesel engines suitable for use in light duty vehicles. Automakers are rapidly commercializing a variety of hybrid electric vehicles, dedicated compressed natural gas vehicles, and flexible-fuel vehicles that run on mixtures containing up to 85 percent ethanol. Fuel cells are being heavily researched and developed. All of these vehicles achieve CO_2 reductions compared to conventional gasoline vehicles. Demand for transportation energy is so immense that no single technology can possibly be the single solution, however.

Anthony Greszler, Vice President of Advanced Engineering at Volvo Powertrain North America, turns to the heavy duty sector. His chapter notes that trucks consume over 20 percent of fuel transportation burned in the United States and that this sector is growing rapidly. He argues that control of CO_2 emissions from heavy duty trucks requires unique metrics, technologies, and public policies. His analysis concludes that it should be possible to achieve 20–30 percent efficiency improvement from proven technologies, but the only realistic way to obtain significant GHG reductions in the face of a growing reliance on heavy duty trucks is to deploy low-carbon alternative fuels. Alternative fuel technologies exist, he adds, but need further development and their cost must be reduced.

Another set of three chapters focus on strategies to tackle transportation GHG emissions not by reducing vehicle emissions, but rather by reducing reliance on automobiles themselves. David Burwell, a Partner in the BBG Group, addresses the question whether or not reducing vehicle miles travelled (VMT) is a sensible strategy for reducing both traffic congestion and transportation-related emissions of CO_2. He finds the answer to be yes, and discusses the leadership of state government agencies in reducing VMT within their jurisdictions.

Rex Burkholder and Eliot Rose from the Portland Metro Council examine the land use and transportation policies that have been successful so far in reducing GHG emissions in metropolitan Portland, Oregon. The Portland metro region has reduced CO_2 emissions, while becoming more liveable and reducing living costs for its residents. The region has implemented a strong land-use planning program that promotes development within an urban growth boundary. This has created a more compact, efficient city that is easier to serve with non-automobile transportation modes. Reliable bus service, streetcar and light rail lines, combined with attention to bicycle and pedestrian planning, ensure that residents who choose not to drive can take advantage of a variety of other travel options.

Gustavo Collantes and Kelly Sims Gallagher from Harvard Kennedy School's Belfer Center for Science and International Affairs modelled individual GHG reduction policies and found that no single policy is likely to achieve meaningful reductions in carbon emissions. One key message is that a policy package—as opposed to an individual policy tool—is necessary to significantly reduce carbon emissions from transportation. They conclude in their chapter that relative stabilization of GHG emissions is likely to be achieved only with a more aggressive taxing scheme. This could induce a meaningful slowdown in VMT increases over time, as well as a stronger adoption of flexible fuel vehicles. As a consequence, they believe oil imports can also be stabilized.

The final two chapters examine the role of consumers in implementing climate change strategies in the transportation sector. Carolyn Fischer, Senior Fellow at Resources for the Future, explores public apprehension over global climate change and its reflection on U.S. fuel economy policy. She argues that the success of the current approach of regulating fuel economy in new vehicles, and hence GHG emissions, depends on whether or not consumers make economically efficient choices. Other approaches, such as a carbon tax on fuel, tolls on roadways, and per-mile charges for driving, may also be needed, she says.

David Greene from the Oak Ridge National Laboratory and his colleagues John German from Honda and Mark A. Delucchi at the Institute for Transportation Studies at the University of California, Davis, explain how markets determine the energy efficiency of durable goods like automobiles. Understanding this is critical to formulating effective policies for mitigating GHG emissions and reducing oil dependence. Their chapter focuses on the consumer trade-off between purchase price and future energy savings of vehicles. The consumer's concern is the net value, the difference between the two. They conclude that this is a risky proposition involving uncertain initial costs and more uncertain future savings. Uncertainties increase the likelihood that loss-averse consumers would decline to bet on new energy efficient equipment even when the expected net present value is positive. They show that typically loss-averse consumers would reject a bet on a fuel economy increase from 28 to 35 miles per gallon, despite an expected present value of about $400 per vehicle.

References

Boarnet, Marlon G. and Randall Crane, *Travel by Design: The Influence of Urban Form on Travel* (New York: Oxford University Press, 2001).

Ewing, Reid; Keith Bartholomew; Steve Winkelman; Jerry Walters; and Don Chen, *Growing Cooler: The Evidence on Urban Development and Climate Change* (Chicago: Urban Land Institute, 2007).

Farrell, Alex; D. Sperling; and others. A Low-Carbon Fuel Standard for California, Part 2: Policy Analysis. Institute of Transportation Studies, University of California, Davis, Research Report UCD-ITS-RR-07-08, 2007.

Handy, Susan; Xinyu Cao; and Patricia L. Mokhtarian, "Self-Selection in the Relationship between the Built Environment and Walking," *Journal of the American Planning Association* 72, no. 1 (2006): 55–74.

Intergovernmental Panel on Climate Change (IPCC), "Climate Change 2007," 2007a accessed from website www.ipcc.ch

Intergovernmental Panel on Climate Change (IPCC), Working Group III Report "Mitigation of Climate Change", 2007b accessed from website www.ipcc.ch

Lutsey, Nic. "Prioritization of Technology Alternatives for Cost-Effective Climate Change Mitigation." Ph.D. Dissertation, Institute of Transportation Studies, University of California, Davis, 2008.

Lutsey, Nic and D. Sperling, "America's Bottom-Up Climate Change Mitigation Policy," Energy Policy, 36: 673–685 (2008).

McKinsey & Company, Reducing U.S. Greenhouse Gas Emissions: How Much at What Cost? Final Report, 2007 http://www.mckinsey.com/clientservice/ccsi/pdf/US_ghg_final_report.pdf

Searchinger, Timothy; Ralph Heimlich; R. A. Houghton; Fengxia Dong; Amani Elobeid; Jacinto Fabiosa; Simla Tokgoz; Dermot Hayes; and Tun-Hsiang Yu, "Use of U.S. Croplands for Biofuels Increases Greenhouse Gases Through Emissions from Land Use Change," Science, February 7, 2008.

United Nations Environment Program (UNEP), "Global Environment Outlook" October 2007, accessed from website www.unep.org

U.S. Energy Information Administration (EIA), "Annual Energy Outlook 2008, " Washington DC: December 2007.

U.S. Environmental Protection Agency (EPA), "Inventory of US Greenhouse Gas Emissions and Sinks: 1990–2005," Washington DC: 2006.

Chapter 2
Energy Security, Climate and Your Car: US Energy Policy and Beyond

Amy Myers Jaffe

The United States (U.S.) is facing daunting energy challenges. Demand for oil has been rising steadily, but growth in supplies has not kept pace. The United States is the third largest oil producer in the world, but its production has been declining since 1970 as older fields have become depleted. It is now more dependent on foreign oil than ever before, importing 12.3 million barrels per day (bpd) in 2006 or about 60 percent of its total consumption of roughly 20.7 million bpd. That is up from 35 percent in 1973. The share of imported oil is projected to rise to close to 70 percent by 2020, with the United States becoming increasingly dependent on Persian Gulf supply. U.S. oil imports from the Persian Gulf are expected to rise from 2.5 million bpd, about 22 percent of its total oil imports, in 2003 to 4.2 million bpd by 2020, at which time the Persian Gulf will supply 30 percent of total U.S. oil imports, according to forecasts by the U.S. Department of Energy's (DOE's) Energy Information Administration (EIA, 2006).

More than three decades after the 1973 oil crisis, U.S. supply of oil is no more secure today than it was thirty years ago. Moreover, its dependence on oil for mobility has never been stronger. All told, there are over 242 million road vehicles in the United States, or one vehicle for every person. Each vehicle is driven over 12,000 miles annually, and virtually all vehicles are powered by petroleum-based fuels, either gasoline or diesel. As a result, despite the fact that the United States accounts for only 5 percent of the world's population, it consumes over 33 percent of all the oil used for road transportation in the world. Future U.S. oil consumption is centered squarely in the transportation sector, which represents more than two thirds of total petroleum use and will constitute over 70 percent of the increase in demand.

As oil demand and dependence on the Middle East rises, the United States has yet to forge a thoughtful response to climate change. In 2005, it emitted a total of 712 million metric tons of carbon, 412 million metric tons of which came from road petroleum use. The country emits more energy related carbon

A.M. Jaffe
Rice University, MS 40, 6100 Main Street, Houston TX 77005, USA

D. Sperling, J.S. Cannon (eds.), *Reducing Climate Impacts in the Transportation Sector*, DOI: 10.1007/978-1-4020-6979-6_2, © Springer Science+Business Media B.V. 2009

dioxide per capita than any other industrial nation (Bryne et al., 2007). In the 1990s, the U.S. transportation sector represented the fastest growing emissions of carbon dioxide than any other major sector of the economy (Romm, 2006). The DOE predicts that the transport sector will generate almost half of the 40 percent rise in U.S. carbon emissions projected for 2025 (EIA, 2006).

The urgent need to reverse the growth path in U.S. fossil fuel use and related global warming pollution has opened debate about the risks and trade-offs of various strategies. There are, in fact, many reasons to be concerned about a major supply disruption that could affect mobility. The United States has no comprehensive strategy to deal with this challenge and perhaps worse still, some of the options available to lessen this risk could come at an expensive cost in terms of climate change mitigation.

This chapter discusses some of the major supply risks that face international oil markets and considers the carbon emission implications of the kinds of energy supplies that countries like the United States may turn to in an effort to diversify away from rising dependence on Middle East oil and to create a more secure energy future.

Rising Demand and Insecure Supply

There are billions of barrels of conventional oil reserves left under the ground and trillions of barrels if more expensive, unconventional resources such as Canadian tar sands are included. But oil is ultimately a finite resource and the geography and geopolitics of bringing oil to the market may render future oil supply less reliable than in the past. Geopolitical conflicts could result in a sudden supply problem or long term problems may also emerge, as future oil supplies fail to materialize in the volumes needed to meet demand.

The security of U.S. energy supply will be highly influenced by international events in the coming years. Geopolitical factors, rather than geology, are likely to drive the energy future. While vast amounts of oil resources remain available for exploitation, more 75 percent of the undiscovered resources outside Organization of Petroleum Exporting Countries (OPEC) are located offshore, according to the U.S. Geological Survey (USGS, 2000). This lends credence to contentions that much of the "easy" oil has already been found. Experts generally agree that world dependence on Middle East oil is likely to grow over time as a natural peak emerges for oil and natural gas production occurs elsewhere. Oil production has already peaked in the United States, for example, and North Sea production, the leading local supplier to Europe, has declined from 6.39 million bpd in 2000 to under 2.11 million bpd at the end 2005. U.S. domestic oil production totaled 5.12 million bpd in 2005, down from 6.48 million barrels a day ten years earlier.

Despite the decline in oil and natural gas production in the United States and Europe, the possibility that the world will geologically "run out" of fossil fuels seems remote. There has been much speculation that future energy challenges will derive from the imminent peaking of the world's existing hydrocarbon resource base. This view seems premature given the enormous resource base that remains across the globe. The World Petroleum Assessment released by the USGS in 2000 estimates that in areas exclusive of the United States, a mean value of 649 billion barrels of oil could be added to recoverable proven reserves through new discoveries through 2025 (USGS, 2000). This estimate is 20 percent higher than the estimate in their last assessment published in 1994.

The USGS also found that total resources, including undiscovered, recoverable resources, reserve additions, remaining proven reserves and cumulative production outside the United States, totaled 2,659 billion barrels, up 5 percent from previously assessed totals. The ratio of global proven reserves to production currently stands at 42 years, substantially higher than it was in 1972.

In additional to these conventional resources, an International Energy Agency (IEA) assessment published in 1998 noted an additional 1.7 billion barrels of unconventional tar sands and oil shale remain to be exploited (IEA WEO, 1998). It is important not to ignore the size of the latter since the potential for producing unconventional resources is improving substantially, given high oil prices and technological progress which has greatly lowered the cost of producing unconventional resources. Canadian unconventional oil production rose by 220,000 bpd in 2007. Tar sands production is averaging about 1 million barrels a day and is expected to rise to 1.7 million bpd by 2010 (EUB, 2006). Gas-to-liquids (GTL) production is also projected to climb in the coming years as companies tap this technology to exploit vast natural gas resources in the Middle East and Africa.

Moreover, the world's proven deposits of coal are plentiful at 984.5 billion tones (IEA, 2001). At current oil prices, even coal-to-liquids conversion can be profitable with inputs of coal at costs of less than $12 a ton, according to Sasol, which operated coal-to-diesel plants in South Africa in the 1980s. Royal Dutch Shell and Sasol are investing in the construction of coal-to-liquids plants in China in a joint venture with Chinese coal companies.

The question, however, is not whether or not there will be enough fossil energy resources under the ground, but whether the geopolitical, social, and environmental factors will hinder the development of these resources. In recent years, political factors have far outweighed geological ones in limiting available supply to world oil markets. In the future, moreover, environmental factors and regulation will increasingly influence what resources get produced. The question will not be are there enough fossil fuel resources under the ground to burn. The question will focus instead on the environmental and legal consequences to burning those resources.

The consequences of continuing to burn fossil fuels at current or expanding rates will worsen global climate changes. Martin Hoffert, professor of physics at New York University and author of a analysis published in *Science* magazine "Advanced Technology Paths to Global Climate Stability: Energy for a Greenhouse Planet," argues that stabilizing the carbon dioxide-induced component of climate change is most fundamentally an energy problem (Hoffert et al., 2002). He noted that stabilization will not only require an effort to reduce end-use energy demand, but also the development of primary energy sources that do not emit carbon dioxide into the atmosphere.

Under a business-as-usual energy supply scenario, carbon concentrations in the atmosphere would rise to 750 parts per million (ppm) by the end of the century, a concentration level Hoffert's calculations say would melt the West Antarctic ice sheets and erode coastlines around the globe. At least 15 terawatts of non-fossil fuel energy will be needed to reduce carbon dioxide levels to modest targets of 550 ppm by 2050. To reach the goal of 350 ppm, at least 30 terawatts would need to be derived from non-fossil sources.

The challenge of restraining demand growth for fossil energy will be monumental. Concerns about resource availability and climate change have been brought to the fore by forecasts of rapidly rising world oil demand in the coming decades. Economic development, in general, is correlated with increased urbanization and electrification, and growth in the use of private automobiles. Research shows that as per capita income rises between $5,000 and $12,000, vehicles stocks per person in a developing nation can increase by as much as a factor of 20. For example, in a country where there are 25 vehicles per 1000 people, as income rises above $5,000 per capita, vehicles stocks will increase to 500 vehicles per 1,000 people (Medlock and Soligo, 2001;2002). This correlation is important because many nations, including China and India, are experiencing per capita income increases to this critical "launching point" for car ownership.

There is a tremendous potential demand for energy use as the global economy expands. Per capita primary energy use in the developing world remains markedly lower than the industrialized West, with India's total primary energy consumption per person averaging roughly 0.38 tonnes of oil per person in 2006, China's at 1.29 and Brazil's at 1.09. These figures are very low compared to the United States average of 7.79 tonnes of oil per person and Germany's 3.98 tonnes of oil per person.

In fact, Asian energy use is expected to expand significantly as the 21st Century progresses. By 2020, Asian energy consumption is projected to account for over one-third of global energy use, rivaling that of North America and Europe and likely resulting in large increases in an already substantial dependence on imported energy. More than half of the future growth in energy demand in Asia is expected to come from the transportation sector where, barring a technological breakthrough, increased reliance on crude oil and crude oil products will be unavoidable. Per capita income growth in developing countries in particular, such as China, Malaysia, Thailand, India, and Indonesia, will account for an increasing proportion of energy demand by encouraging an increase in automobile ownership, and with it, a corresponding rise in motor fuel demand (Medlock and Soligo, 1999; 2001; 2002).

To put this into perspective, total oil demand for the region is already larger than that of the United States, and oil imports, which are already above 70 percent of total consumption, have risen substantially in recent years, up from about 11 million bpd in 1998 (EIG, 2001). According to "business as usual" scenario forecast by the IEA, oil demand in all of Asia is expected to grow two to three times faster than in the industrialized West. By 2010, total Asian oil consumption could reach 25–30 million bpd (IEA, 2000).

Global oil demand is expected to rise at a rate of roughly 1.6 percent per annum over the next two decades from about 76.4 million bpd in 2001 to 90.4 million bpd in 2010 and 106.7 million bpd by 2020 (IEA, 2004). Almost 75 percent of this increase in demand will come from the transport sector where renewable energy and nuclear energy are not expected to play a significant role without a major technological breakthrough.

Under a "business as usual" scenario, much of this increased demand for oil, roughly 60 percent, will have to be supplied by rising production from OPEC over the next 25 years (IEA, 2005). The reality of conventional oil and natural gas geology is that approximately 62 percent of remaining proven resources lie in only five countries. In the case of oil, the five largest resource holders are all Middle Eastern countries. In projecting future supply potential, more than half of that volume is projected to come from just three countries shown in Fig. 2.1: Iraq, Iran, and Saudi Arabia. These forecasts might prove unrealistic given the

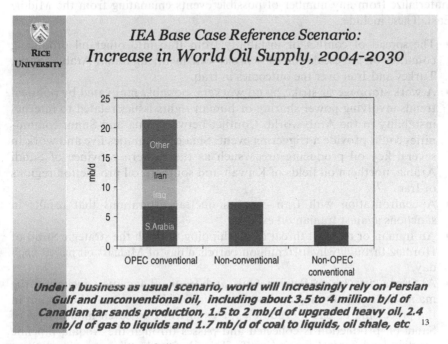

Under a business as usual scenario, world will increasingly rely on Persian Gulf and unconventional oil, including about 3.5 to 4 million b/d of Canadian tar sands production, 1.5 to 2 mb/d of upgraded heavy oil, 2.4 mb/d of gas to liquids and 1.7 mb/d of coal to liquids, oil shale, etc [13]

Fig. 2.1 Future new supplies of oil

political and economic conditions in those countries. Conventional oil production from non-OPEC countries is expected to play a markedly smaller role by providing just 10–15 percent of increased supply.

This means that nonconventional resources will play an increasingly important marginal supply role by supplying 25–30 percent of future oil supplies. Canadian tar sands production would represent the largest possible diversity away from Middle East supply at over 3.5–4 million bpd. Upgraded heavy oil could yield another 1.5–2 million bpd, while an additional 1.7 million bpd of production expected from coal-to-liquids and oil shale production. Gas-to-liquids output is expected to reach 2.0–2.5 million bpd. Without the development of these unconventional resources, the world will be even more dependent on Middle East supply. However, the pursuit of these unconventional resources is not without a downside. They all tend to have a higher carbon footprint, pitting energy security goals against climate priorities.

Risks to Middle East Oil

The need to diversify the heavy dependence on Saudi, Iraq and Iranian oil in the United States is driven home by the tensions and conflict that now plague the Persian Gulf. Many Persian Gulf nations currently face both internal instability and future succession problems. A severe oil supply shock could potentially materialize from any number of possible events emanating from the Middle East. These include:

- The spread of conflict or instability from Iraq into other oil producing countries or the escalation of a proxy war involving Saudi Arabia, Syria, Turkey and Iran over the outcomes in Iraq.
- A work stoppage or strike by oil workers, possibly motivated by political trends involving power sharing or human rights issues related to internal instability in the Arab world. Conflict between Shia and Sunni communities could provide a triggering event. Shia communities live and work in several key oil producing areas such as the Eastern province of Saudi Arabia, northern oil fields of Kuwait and southern oil production regions of Iraq.
- A confrontation with Iran over its nuclear aspirations that results in sanctions against Iranian oil exports.
- An Iranian or terrorist threat to oil shipping through the strategic Strait of Hormuz through which 16 million barrels a day of Mideast oil passes each day.
- Al-Qaeda attacks on Persian Gulf and Iraqi oil facilities, including the main Saudi oil gathering center of Abqaiq which is a critical chokepoint in the processing of Saudi crude oil for export. Six to seven million bpd of Saudi oil production is collected and passes through the Abqaiq pipeline junction and central processing facilities before heading to export from

Saudi loading terminals at Ras Tanura and Ju'aymah. A loss of access to Saudi oil production facilities at Abqaiq could leave world markets suddenly with a major supply deficit. Under certain damage scenarios, certain facilities could not be easily fully replaced in the three month timeframe where strategic stocks could be used to make up for lost Saudi barrels.

- An exodus of oil workers occasioned by fear of terrorism, domestic unrest, or swift change in ruling regime.
- Domestic unrest or political crises, ranging from a leadership succession problem to a radical revolutionary challenge to an existing regime.

An expanded proxy war in Iraq fanned by the actions of its neighbors could create a crisis of even greater proportions than currently seen inside Iraq's borders and would be detrimental to the region as a whole. An expansion in violence in Iraq and beyond would greatly damage the stability of the oil market.

The United States and its allies have faced uncertain times in the Persian Gulf before, beginning with the 1956 Suez crisis and extending into the 1980s and 1990s with the Iranian revolution, the eight year war between Iraq and Iran, and the Gulf war. What makes today's situation more challenging is the fact oil refining capacity, and oil transportation capability are facing acute capacity shortages today. Any disruption of Middle East oil flows as a result of a conflict between countries in the Persian Gulf region would have deep and instantaneous ripple effects into the market not seen in earlier crises. Fears of such scenarios are one explanation for oil prices approaching $100 a barrel in November 2007 amid official threats of a Turkish military incursion into northern Iraq against Kurdish militias and diplomatic confrontations between the United States and Iran over Tehran's plans to pursue nuclear capability.

Saudi Foreign Minister Prince Saudi al-Faisal bin Abdul Aziz has raised the specter of the conflict in Iraq becoming a proxy war between Gulf Sunnis and Shi'as, potentially engulfing the entire region, including Saudi Arabia, Iran, Syria, and Turkey (Baker Institute, 2005). Noted the minister:

> The real danger is in the division that is being projected between the Arabs of Iraq, dividing them into Shias and Sunnis, especially a separate entity for both... This is a recipe for bringing the countries around Iraq into conflict themselves. You have Iran on one side which will come in with the Shias. We have the Turks on the other side which will come in to fight with the Kurds, and the Arabs will definitely be dragged into the fight on the part of the Sunnis... Unless the Sunnis and Shias are brought together, it will disintegrate into civil war... and then, the whole region will also disintegrate and conflicts that we have not dreamt of in the past will be facing the international community.

The Saudi government has as strong interest in national reconciliation in Iraq and for the peaceful coexistence of Sunni and Shia Arab populations. With the rise of a Shi'a-dominated government in Baghdad, Iran has been able to expand its influence in Iraq, a development of concern to Saudi Arabia and other countries with regional Arab Sunni majorities. With its own Shi'a minority estimated by some to be between 10 and 20 percent of its population, Saudi Arabia clearly worried about a "pan-Shi'a" movement in the Persian Gulf hostile to the Saudi regime. The possibility of popular unrest in Shi'a areas is no small matter of concern for Riyadh. The majority of Saudi Shi'as lives in the oil-rich Eastern province where the vast bulk of Saudi Arabia's oil production is located. A majority of skilled workers for Saudi Aramco, the state oil mono-poly, in the Eastern province oil fields are of Shi'a origin despite a program to diversify the workforce in recent years. This means any kind of politically-motivated work stoppage, strike, social protest or repressive clamp down could have immediate ramifications for stable oil production flows (Jaffe and Barnes, 2006).

Hints that Saudi Arabia might back Sunni fighters inside Iraq to protect its interests against Iranian-backed militias are a warning of possible negative scenarios that could emerge if stability cannot be achieved in Iraq through political means. For its part, Iran has put its Gulf neighbors on notice that it could be more aggressive, with Hussain Shariatmadari, an advisor to Iranian Supreme Leader Ali Khamenei and managing editor of the Iranian daily *Kayhan*, claiming recently that Shia populations in Bahrain demand the reuni-fication of "this province of Iran to its motherland, the Islamic Republic of Iran." He added, "It goès without saying that such an indisputable right for Iran and the people of this province should not and cannot be overlooked."(*World Tribune*, 2007)

In addition, Iranian assertiveness on its pursuit of nuclear capabilities and support for terrorism has unsettled oil markets with worries that the United States and its allies will impose sanctions against Iran or that Tehran will suspend its oil exports as a protest against any United Nations action against it. Tehran has geographical leverage on the flows via the Strait, and Iran's pursuit of nuclear capability must be seen in this light (Dagobert and Jaffe, 2005). Tensions between Iran and the West and between Iran and Israel remain a feature driving price volatility today. Israeli officials have warned that Tel Aviv could hit vulnerable oil export facilities like Kharg Island and other offshore facilities instead of preemptively attacking the Bushehr nuclear plant, an event that could prompt Tehran to escalate by retaliating against other regional oil facilities, as discussed by Geoff Kemp of the Nixon Center in his monograph "U.S. and Iran: The Nuclear Dilemma, Next Steps" (Kemp, 2004). If Iran felt its own oil export capability was about to be or had been destroyed, it might counter by issuing a threat to damage oil facilities or block exports from other nearby U.S. allies. Israel's conflict with Hizbollah in Lebanon in the summer of 2006 highlighted the dangers of lingering conflict that could, if not properly managed by effective diplomacy, expand to embroil a wider range of countries—including Syria and Iran—where active support for such subna-tional groups as Hamas and Hizbollah continues to be a destabilizing factor.

Beyond inter-nation conflict scenarios, stable oil supply from the Persian Gulf is also threatened by problems related to internal instability and terrorism. Such problems simmer against the backdrop of the large U.S. military presence in the region and a rising tide of anti-American sentiment, linked in part to regional perceptions about U.S. oil lust. A 2004 poll by Pew Research Center found that:

> In the predominantly Muslim countries surveyed, anger toward the United States remains pervasive, although the level of hatred has eased somewhat and support for the war on terrorism has inched up. Osama bin Laden, however, is viewed favorably by roughly half of the survey respondents in Pakistan, Jordan and Morocco. Majorities in all four Muslim nations surveyed doubt the sincerity of the war on terrorism. Instead, most say it is an effort to control Mideast oil and to dominate the world (Pew Research Center, 2004).

In 1995, al-Qaeda leader Osama Bin Laden wrote a letter to the late King Fahd of Saudi Arabia, enumerating his grievances against the governments of Saudi Arabia and the United States. In the communication, made public to his followers, Bin Laden sharply criticized the servitude of Saudi oil policy to U.S. interests and decried the loss of oil revenue the kingdom suffered as a result of these policies. He argued that since the real price of oil was held artificially low, the Muslim world lost $36 trillion over a quarter century, or $30,000 for each of the world's 1.2 billion Muslims. In his letter to the American people, Bin Laden summed up popular sentiment on the subject: "You steal our wealth and oil at paltry prices because of your international influence and military threats. This theft is indeed the biggest theft ever witnessed by mankind in the history of the world" (Global Security, 2006).

Despite its grievances related to oil, Bin Laden and other Islamic radicals believed until recently that energy facilities in the Muslim world should be spared, since they constituted the wealth of a future Islamic state. By December 2004, however, Bin Laden changed his mind and called for attacks on oil facilities as part of the jihad against the West. Al-Qaeda unsuccessfully tried to attack the major crude oil processing facilities at Abqaiq in February 2006. In a message claiming responsibility for the attack, Al-Qaeda of the Arabian Peninsula said the attack was part of the war against "Christians and Jews to stop their pillage of Muslim riches" (Daly 2006).

Internal instability has also thwarted Iraq from developing its massive oil resources. Iraq's oil sector is in even more disarray now than before the U.S. invasion. Strategies for increased oil development in Iraq are politically contentious as Iraq attempts to overcome sectarian and ethnic differences given the unstable security environment inside the country and the concentration of major oil resources in Shi'a controlled regions. The politics of deciding who controls future oil development is complicating Iraq's ability to forge a permanent government structure with full federal authority. Instead, regional leaders are trying to assert their optimum control over the future cash flow to be generated by oil production, hindering political compromise and national unity. Between this political uncertainty and the poor security situation on the ground, there is a high probability that major production increases will not occur in Iraq for several years, if ever.

Over the past two decades, the U.S. oil policy has been to rely on allies in the Persian Gulf, such as Saudi Arabia, the United Arab Emirates, Kuwait, Qatar and Oman, as well as major exporters like Venezuela and Nigeria to provide oil. In 1990, when Iraq invaded Kuwait, cutting off 5.0 million bpd of oil supply, Saudi Arabia, the United Arab Emirates, Nigeria, and Venezuela increased production to make up the difference, limiting the effect on world oil supply and price.

But the internal stability of many of these large oil producing countries looks a lot shakier now than it did in the 1980s and 1990s. In fact, the list of oil exporting countries where production has been stagnant or falling in recent years despite ample reserves due to civil unrest, terrorism, inefficiency, government mismanagement, or corruption is long and diverse. Projections that OPEC will increase capacity by an additional 10–20 million bpd in the next 20 years to meet the rising demand discussed above run counter to historical experience. OPEC's capacity has fallen, not increased, over the past 25 years, from 38.76 million bpd in 1979 to roughly 31 million bpd today.

Many factors have contributed to OPEC's inability to expand its sustainable oil production capacity. In the late 1980s, OPEC had planned to increase its oil field production capacity to 32.95 million bpd by the mid-1990s. Instead, OPEC production capacity stagnated at 29 million bpd for most of the decade, only creeping higher in recent years. Even so, large capacity expansion programs in Saudi Arabia, Iran, Libya, and Iraq have all failed to achieve production targets due to international sanctions. Venezuela's planned expansions were thwarted by a change of government, related civil unrest and a redirecting of funds away from the oil sector to social welfare programs, and the country's oil potential has been slipping in recent years. Regional and ethnic conflict and civil unrest also plagued Nigeria's efforts to expand production, while domestic politics has blocked oil field investment in Kuwait.

National Oil Companies

Unlike past decades when private, publicly traded oil companies played a major role in the worldwide oil exploration business, national oil companies (NOCs) will be responsible for the lion's share of the increase in oil output and investment in the next twenty years. State-owned NOCs represent the top oil reserve holders internationally. In 2005, NOCs controlled 77 percent of the global proved oil reserves and partially or fully privatized Russian oil companies controlled another 6 percent. These government-controlled companies do not allow equity participation by foreign oil companies. By comparison, western international oil companies (IOCs) that dominated the oil scene in the 20th Century now control less than 10 percent of the world's oil and gas resource base.

NOCs will overwhelmingly dominate world oil investment, production and pricing in the coming decades. As the world becomes more dependent on NOCs for future oil supplies, major oil consuming countries are questioning the ability of these firms to bring on line new oil in a timely manner in the volumes that will be needed, stimulating new debate about long term energy security.

The list of NOCs with falling or stagnant oil production in recent years due to civil unrest, government interference, corruption and inefficiency, and the diversion of corporate NOC capital to social welfare is long. It includes a wide range of oil-rich countries, such as Indonesia, Iran, Iraq, Mexico, Russia, and Venezuela. To the extent that NOCs must meet national socio-economic obligations, such as income redistribution, over-employment, fuel price subsidization, and industrial development, they have fewer incentives or resources for reinvestment, reserve replacement, and sustained exploration and production activity.

This raises the question whether timely development of the vast resources under the control of NOCs can take place given the constraints imposed by domestic political influences and geopolitical factors. The tendency of NOCs to focus on socio-economic activities other than oil field maintenance and expansion is partly responsible for the slow pace of resource development relative to the rapid rise in global demand and could mean that new production will not materialize to meet rising oil requirements in the future, leaving major oil consuming nations scarce of fuel (Baker Institute, 2007).

Weather-Related Issues

Tensions in the Middle East and the rising control of NOCs are not the only risks facing U.S. oil supply and with it, transportation mobility. The U.S. fuel system remains vulnerable to severe storms and other weather related disruptions. National energy prices spiked in the wake of supply disruptions caused by hurricanes along the U.S. Gulf Coast coastline during the summer and autumn of 2005. For the first time in three decades, drivers found "no gas" signs on gasoline pumps across the Gulf coast and up the eastern seaboard. By September 1, 2005, New York Harbor gasoline spot prices rose to just over $3.00 per gallon, up by $1.16 per gallon prior to the hurricane. Average retail prices peaked at $3.12 per gallon (AAA Fuel Gauge Report, 2005). As shown in Fig. 2.2, the spikes experienced by gasoline were higher and shaper than the increases in heating oil, showing the vulnerability of transportation fuels to supply disruptions.

In the aftermath of Hurricane Katrina, Gulf coast refinery production of finished gasoline fell by 700,000 bpd versus year earlier levels. Hurricane Rita made landfall in Texas on September 24, 2005, and resulted in an additional, larger loss of refining capability. For the week ending September 30, finished gasoline production was down by 1.4 million bpd versus levels a year earlier.

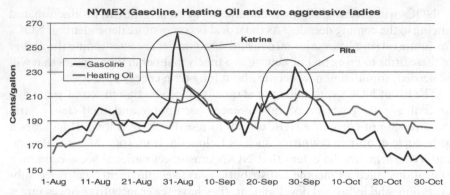

Fig. 2.2 NYMEX gasoline and heating oil price spikes after two hurricanes

Seventy-five days after the hurricanes, over 90 million barrels of crude oil and over 175 million barrels of refined products had been lost from the market. In December 2005, close to 750,000 bpd of U.S. refining capacity was still affected by the aftermath of the hurricanes and was not brought back on line until end of March 2006.

The hurricanes of 2005 affected not only refinery capacity, but also negatively influenced deliverability of product. For example, the immediate aftermath of Hurricane Katrina, which forced the shutdown of two main gasoline transport pipelines from the Gulf coast to the Eastern seaboard, created temporary fuel shortages at retail stations from Florida all the way to Canada. Retail gasoline prices reached as high as $6.99 per gallon were reached in some markets.

The events of 2005 highlighted the possible dangers of having so much U.S. refining capacity concentrated in one geographical region that is vulnerable to weather-related disruptions. The area stretching from Corpus Christi, Texas, to Lake Charles, Louisiana, is home to 21 refineries, comprising 27 percent of U.S. refining capacity. The Houston/Beaumont/Port Arthur area of Texas represents 20 percent of U.S. refining capacity. The Gulf of Mexico provides 29 percent of U.S. domestic crude oil production and 19 percent of its domestic natural gas supply. This heavy geographic concentration of oil refining and energy production means that similar or worse disruptions are possible in the future, especially if global warming and sea level rise contribute to an escalation of severe weather along the U.S. Gulf coast.

The problem is compounded by the fact that U.S. refinery utilization has averaged above 90 percent for several years. Operation of the energy system at levels so close to capacity limits means that unexpected outages can quickly lead to gasoline price spikes and even regional physical shortages. U.S. refineries produce about 8.8 million bpd of gasoline on average and roughly 9.2 million bpd during the peak production season in the spring and summer, just before the hurricane season. The summer demand peak for gasoline in 2005 was 9.7 million bpd, and this gap between refinery output and consumer demand was met by imports. This trend has increasingly contributed to summer price spikes and leaves the United States even more vulnerable to weather-related refinery problems.

Policy Options: Meeting the Challenge of Rising Oil Demand

In his 2007 State of the Union address, President G.W. Bush announced an ambitious target to reduce the growth in U.S. gasoline use by 20 percent over the next ten years. The president noted that the nation was "addicted to oil" and added that U.S. dependence on imported oil makes it "more vulnerable to hostile regimes, and to terrorists—who could cause huge disruptions of oil shipments, raise the price of oil, and do great harm to our economy." The president outlined his program by proposing to increase the supply of renewable and alternative fuels by setting "mandatory fuels standards" to require 35 billion gallons of renewable and alternative fuels in 2017, roughly displacing 15 percent of projected annual gasoline use in that year. The president's plan also called for modernization of the corporate average fuel economy (CAFE) standards and rules for light trucks to reduce projected annual gasoline use by 8.5 billion gallons, representing a future 5 percent reduction in gasoline demand.

The President's proposal supplementing the DOE target goal under the Energy Policy Act of 2005 requires that 30 percent of 2004 U.S. transportation fuel consumption be displaced with biofuels by 2030. Renewable and alternative fuels were defined by the White House as corn ethanol, cellulosic ethanol, biodiesel, methanol, butanol, hydrogen, and alternative fuels (White House, 2007).

The U.S. Congress responded with even stronger legislation. The Energy Independence and Security Act of 2007, passed on December 18, 2007 and signed by President George W. Bush, raised automobile CAFE standards to 35 miles per gallon (mpg) by 2020, with first improvements required in passenger fleets by 2011. The legislation also increases the Renewable Fuels Standard (RFS) to require nine billion gallons of renewable fuels consumed annually by 2008 and progressively increase to a 36 billion gallon renewable fuels annual target by 2022, of which 16 billion is slated to come from cellulosic ethanol. The law specifies that 21 billion gallons of the 36 billion must be "advanced biofuel" which has 50 percent less greenhouse gas (GHG) emissions than the gasoline or diesel fuel it will replace. "Advanced biofuels" include ethanol fuel made from cellulosic materials, hemicellulose, lignin, sugar, noncorn starch and wastes, and biomass based biodiesel, biogas, and other fuels made from cellulosic biomass.

The 2007 Energy Independence and Security Act is the first serious national energy legislation passed in decades aimed to achieve even the modest conservative goal of holding gasoline demand flat between 2005 and 2017. To meet even this goal will require U.S. ethanol production to increase seven-fold, or 16 percent per year for 10 years. While ethanol production did increase by this percentage in 2006, continuing to grow at this pace is likely to be a challenge. In fact, current levels of ethanol production have already led to increases in corn-based food prices, and analysts worry that in drought conditions, including dry conditions that might occur in the Midwest due to global warming trends, the consequences of ethanol production on food costs could be severe. Moreover, some studies show that the environmental impact of increased use of fertilizers and irrigation use on ecosystems along the Mississippi River and in the Gulf of Mexico could be drastic (Powers et al., 2001; NREL, 2005).

The law also failed to close the loophole on how flexible fuel vehicles (FFVs) are treated under the CAFE standards. Under the current fuel economy regulations, car manufacturers receive a 1.2 mpg credit for every FFV that they produce that can run on either conventional gasoline or a gasoline mixture containing up to 85 percent ethanol (E85). This is problematic, though, because few of the FFVs being used to earn this credit for automakers actually operate on alternative fuel. Less than 1,000 out of 170,000 gasoline stations in the United States offer E85 fuel. Current U.S. ethanol production is concentrated in the Midwest region and difficulties remain with the distribution system to other parts of the country. Of the 1,300 E85 ethanol fueling stations in the United States in December 2007, over 35 percent of them were located in two states, Minnesota and Illinois (DOE, 2007).

In recent years, higher demand growth for transportation fuel has reflected growing per capita income, demographic growth and a trend towards larger cars. The future growth in U.S. oil use will come mainly from the transportation sector. To address reductions in oil use, it will be necessary to address the efficiency issue for motor vehicles. Many experts say that higher gasoline prices have not "mattered" to American drivers and rising prices are not likely to affect demand either. This has not actually been the case, however. Prior to 2006, U.S. gasoline demand had grown almost every year except during the 1991 recession and in the aftermath of the September 11, 2001, attacks. On average, demand grew by 1.7 percent per annum between 1985 and 2004. In 2005, U.S. gasoline use rose by 1.2 percent but in 2006, gasoline use was down 0.3 percent over year ago levels, in the face of rising pump prices.

Even to hold U.S. gasoline use at 2005 levels by 2017 will require a 25 percent improvement in per vehicle fuel economy just to counteract the projected increase amount of driving. The fuel economy of new cars will have to increase even more given the large number of older and less fuel efficient vehicles that will remain on the road. According to a Department of Energy study, 75 percent of all cars remain in circulation at least 10 years (OTT, 1998). Thus, it will take more than a decade for higher standards to dramatically alter the average performance of the on road efficiency of the entire U.S. car fleet.

If about 20 percent of the cars on the road are replaced by new vehicles over the next ten years—a pace in line with the past decade—all new cars sold between now and 2017 would have to average 42 mpg, far higher than the 35 mpg targeted in the 2007 Energy Independence and Security Act for 2020. If prices or tax credits pushed consumers to replace 50 percent of all cars on the road—a pace in line with rates seen in the late 1970s and early 1980s—the average on road efficiency of new cars would only have to be 26.5 mpg (Rice, 2007). Each 1 percent improvement in fuel economy across the entire U.S. vehicle fleet saves close to 600,000 barrels of oil imports per day. Additional efficiency gains would save even more oil, but the savings diminish as better mileage performance tends to cause a rebound effect by promoting increased driving,

These back-of-the-envelope calculations make it clear that talking of U.S. energy independence is ridiculous. Eliminating 12 million bpd of oil imports is not plausible. To achieve oil independence by replacing gasoline with ethanol would require approximately 10 times the current amount of worldwide biofuels production. Energy conservation through increased use of public transportation or other means has a role to play, but these measures are also unlikely to achieve substantial fuel savings. To hold U.S. gasoline demand at 2005 levels by 2017, each vehicle would have to be driven about 45 miles less per week by 2017 (Rice, 2007).

Climate Implications

Emissions from the burning of gasoline and other liquid fuels constitute more than one-third of all global emissions stemming from fossil fuel combustion, as shown in Fig. 2.3. Thus, addressing the fuel efficiency issue or reducing automobile use would be an effective means to lower greenhouse gas emissions.

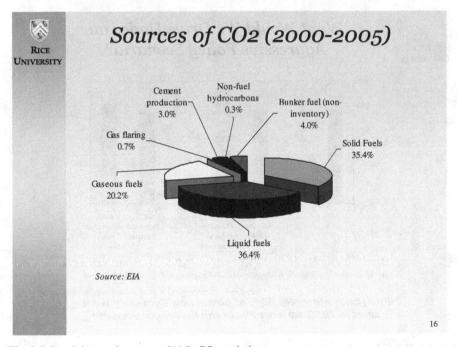

Fig. 2.3 Breakdown of sources of U.S. CO_2 emissions

Joseph Romm notes that "The urgent need to reverse the business as usual growth path in global warming pollution in the next two decades to avoid serious if not catastrophic climate change necessitates action to make our vehicles far less polluting" (Romm, 2006). He believes the most cost effective strategy in the near term is to improve vehicle efficiency, mainly through the use of gasoline burning hybrid electric vehicles. A doubling of onroad efficiency to just over 35 mpg by 2030, possible with widespread deployment of hybrids, would reduce U.S. oil use by 1.6 million bpd and reduce carbon emissions from 542 metric tons to 442 metric tons per year.

Longer term, Romm argues that the United States will have to replace gasoline with zero carbon fuels, a shift that will take strong government action to succeed. Romm notes that hydrogen may prove "the most challenging of all alternative fuels" and argues that advanced plug-in hybrid electric vehicles that would run partially on renewable energy might provide a more promising pathway.

It will be hard to reduce carbon emissions through improvements in car efficiency alone. The IEA projects that under an aggressive carbon policy initiative scenario, a less ambitious form of hybrid electric vehicle known as the mild hybrid could represent up to 60 percent of global new light-duty vehicle sales by 2030, up from roughly 7 percent in the business as usual scenario. This is shown in Fig. 2.4.

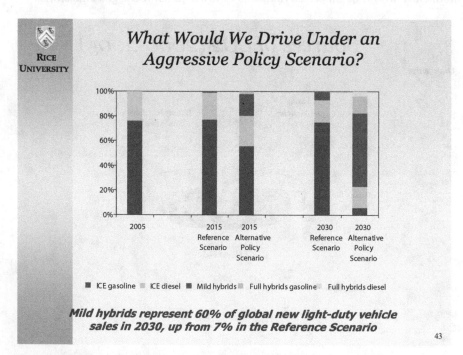

Fig. 2.4 IEA vehicle projections under an aggressive policy scenario

Under an aggressive policy response undertaken scenario, the IEA says that 6.3 gigatons, or 16 percent of energy related carbon, could be reduced, including 10 percent through increased nuclear energy use; 12 percent increased renewable energy use; 13 percent through power sector efficiency improvements; 20 percent through electricity end use consumer efficiency; and 36 percent through transportation and other fuel use efficiency gains (IEA, 2006). These projections are shown graphically in Fig. 2.5.

The state of California is leading the United States with GHG emissions legislation that would reduce the state's emissions at 1990 levels by 2020. Mandatory caps for large emitters, including businesses, utilities and industry, will begin in 2012. The state is also implementing a low-carbon fuel standard that is designed to stimulate improvements in transportation fuel technologies and enhance increased competition among transportation fuels (UCD, 2007). The standard will apply to all fuels sold in California and involve a uniform statewide baseline that will gradually require a decline in carbon intensity. The standard aims to reduce the carbon intensity of California's passenger vehicles by 10 percent in 2020 based on full life-cycle carbon fuel cycle analysis. The regulations will apply to entities that produce or import transportation fuel for use in the state.

Detractors argue that this system will press fuel providers to choose fuels based only on carbon content and cost, while ignoring questionable environmental impacts, for example, deforestation by Brazilian ethanol producers. The

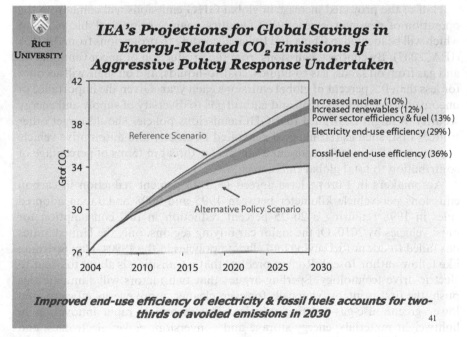

Fig. 2.5 IEA projections of global CO_2 savings under aggressive policy scenario

low-carbon fuel standard has also been cited as possibly discouraging the development of Canadian tar sands, which is more carbon intensive in its mining than traditional onshore oil and natural gas production, but could have an important role to play in the diversity of supplies from the Middle East. Tar sands production results in 0.09–0.16 tons of carbon emissions per barrel produced, or roughly 167 million tons of carbon per year from an output of 1 million bpd.

Another target for a low-carbon fuel standard might be GHG emissions from the flaring of natural gas during traditional oil production. Flaring currently contributes about 400 million tons of carbon a year, over three times the level projected for tar sands, and it poses an immediate health risk to local populations. Flaring represents the same scale of emissions from all vehicles in the United Kingdom, France and Germany, for example. It could be easily banned through national legislations or an international climate initiative. Some oil companies, notably Shell and BP, have moved to reduce flaring in their operations and many countries, such as Algeria and Nigeria, have been pressing foreign operators to end flaring and to capture and use it to provide fuel for local power generation and industry. Countries whose oil production represents the leading contributors to global gas flaring include Nigeria, Russia, Iran, Algeria, Mexico, Venezuela, Indonesia and the United States. From an energy security point of view, an end to flaring requires no energy security tradeoff. Increased capture and sale of natural gas could potentially add to diversity of supply, rather than decrease it.

Half of the projected increase in global GHG emissions will come from the operation of new power generation facilities, mainly using coal and many of which will be located in China and India, according to projections from the IEA (IEA, 2007). By contrast, emissions from the production of unconventional oil and gas from oil sands, gas-to-liquids, coal-to-liquids, and oil shale will account for less than 0.3 percent of global emissions each year. Given the importance of unconventional sources of oil and natural gas to diversity of supply and energy security, it might be argued that GHG emissions policies should target other sources first, such as flaring, coal-generated electricity and automotive vehicle operations, all of which represent a far greater threat in terms of percentage of contribution to total global emissions.

Automakers in Europe have agreed to a 25 percent reduction in carbon emissions per vehicle kilometer between 1995 and 2008 and Japan adopted rules in 1998 requiring a 20–25 percent reduction in fuel consumption for most vehicles by 2010. Of the major car buying regions, only the United States has failed to adopt fuel saving and climate policies in the 1990s. Dan Sperling, like fellow author Joseph Romm, predicts that a transition is about to occur to electric-drive technology. Sperling argues that two factors will stimulate this push: the first is intensifying calls for even cleaner, more energy efficient, and lower greenhouse-gas-emitting vehicles. The second is rapid innovation in lightweight materials, energy storage and conversion, power electronics, and computing (Sperling 2003).

In tackling the reduction of GHG emissions in a manner that least interferes with the promotion of energy security, new policies should focus on how to promote electric-drive technology and hybrid electric vehicle technology to play an increasingly important role rather than barring nonconventional energy sources that will help diversify the energy supply from dependence on supplies from the volatile Middle East.

References

AAA Fuel Gauge Report, September 2005 available at http://www.fuelgaugereport.com/
Alberta Energy and Utilities Board (EUB), 2006, http://www.gov.ab.ca/
The Baker Institute, Transcript of Lecture Address, www.bakerinstitute.org/events_ sept26_-transcript.pdf, 2005
Baker Institute study, The Changing Role of National Oil Companies on International Energy Markets, Baker Institute, 2007, available at www.rice.edu
Brito, Dagobert and Amy Myers Jaffe, "Reducing Vulnerability of the Strait of Hormuz" *Getting Ready for a Nuclear-Ready Iran*, published by the U.S. Army War College, November 2005
Byrne, John, Kristen Hughes, Wilson Rickerson, and Lado Kurdgelashvili, "American Policy Conflict in the Greenhouse: Divergent Trends in Federal, State, and Local Green Energy and Climate Change Policy" Energy Policy, Vol. 35, 4555–4573, 2007
Daly, John C, "Saudi Oil Facilities: Al-Qaeda's Next Target?" Terrorism Monitor, Middle East Institute, Vol. 4, Issue 4, February 2006 as published on http://www.saudi-us-relations.org/articles/2006/ioi/060224-daly-saudi-target.html
Energy Intelligence Group (EIG) Oil Market Intelligence data base, by subscription, July 2001
Globalsecurity.org, Text of Osama Bin Laden "Letter to the American People" available at http://www.saudi-us-relations.org/articles/2006/ioi/060224-daly-saudi-target.html
International Energy Agency (IEA), World Energy Outlook, Paris, France, various years (1998, 2000, 2001, 2004, 2005, 2006, 2007)
Hoffert, Martin, Ken Caldeira, Gregory Benford, et al., "Advanced Technology Paths to Global Climate Stability: Energy for a Greenhouse Planet" Science Magazine, Vol. 298, No. 5595, 981–987, November 1, 2002
Jaffe, Amy Myers and Joe Barnes, "The Persian Gulf and the Geopolitics of Oil, IISS Survival, Vol. 48, No. 1, Spring 2006
Kemp, Geoffrey, U.S. and Iran: The Nuclear Dilemma: Next Steps, available at www.nixoncenter.org, April 1, 2004
Medlock, Kenneth and Ron Soligo, "The Composition and Growth in Energy Supply in China." Houston: The Baker Institute for Public Policy. 1999
Medlock, Kenneth and Ronald Soligo, "Economic Development and End-Use Energy Demand" The Energy Journal, Vol. 22, No. 2, 2001
Medlock, Kenneth and Ronald Soligo, "Automobile Ownership and Economic Development – Forecasting Motor Vehicle Stocks to 2015" The Journal of Transport Economics and Policy, Vol. 36, No. 2, 163–188, Spring 2002
National Renewable Energy Laboratory (NREL). "Quantifying Cradle to Grave Life Cycle Impacts associated with Fertilizer used by Corn, soybeans and stover production," NREL/TP-510-35700, May 2005
Pew Center Global Attitudes Project, "A Year After Iraq War: Mistrust of America in Europe even higher, Muslim anger persists" Pew Research Center, 2004 Available at http://people-press.org/

Powers, Susan, Pedro J.J. Alvarez, and D. Rice, "Increased Use of Ethanol in Gasoline and Potential Groundwater Impacts," State of Calf. UCRL-AR-145380, 2001

Rice University, "Gas FAQ: U.S. Gasoline Markets and U.S. Oil Import Dependence, U.S. Government Policy Analysis Report, www.rice.edu/energy/publications/FAQs/WWT_FAQ_gas.pdf 2007

Romm, Joseph, "The Car and Fuel of the Future" Energy Policy, Vol. 24, 2609–2614, 2006

Sperling, Daniel. "Cleaner Vehicles" Handbook of Transport and the Environment, Ed. D.A. Hensher and K.J. Button. Amsterdam: Elselvier Ltd., 2003

U.S. Department of Energy, Alternative Fuels and Advanced Vehicles Data Center: "Alternative Fueling Station Total Counts by State and Fuel Type." http://www.eere. energy.gov/afdc/fuels/stations_counts.html (Dec.06, 2007)

U.S. Department of Interior, U.S. Geological Survey (USGS), U.S. Geological Survey World Petroleum Assessment, 2000, available at http://pubs.usgs.gov/dds/dds-060/

U.S. Energy Information Agency (EIA), Annual Energy Outlook, 2005, U.S. Department of Energy, Washington, DC, 2006

U.S. Office of Transportation Technologies (OTT), Office of Energy Efficiency and Renewable Energy, U.S. Department of Energy, "The Impacts of Increased Diesel Penetration in the Transportation Sector" Report Number SR/OIAF/98-02, 1998, available at http://www.eia.goe.gov/oiaf/servicerpt/intro.html#ste

University of California-Davis (UCD), "A Low Carbon Fuel Standard for California: Part 2, Policy Analysis, UCD-ITS-RR-07-08 2007 available at http://its.ucdavis.edu

The White House, http://www.whitehouse.gov.stateoftheunion/2007/initiatives/print/energy. html, 2007

World Tribune.com, "Iran stakes claim to Bahrain: Public Seeks 'Reunification'... with its motherland" Friday July 13, 2007, available at http://www.worldtribune.com/worldtribune/WTARC/2007/me_iran_07_13.asp

Chapter 3
Transport Policy and Climate Change

Jack Short, Kurt Van Dender and Philippe Crist

Actions to combat climate change are now at unprecedented levels. With transportation, however, the gap between political aspirations and trends is widening. Achieving greenhouse gas (GHG) reductions presents a major challenge for the transport sector, different in nature and degree from other challenges like reducing accident rates or emissions of traditional pollutants.

This chapter documents the current and expected future importance of transport as a source of GHG emissions. It then examines present policy measures, with a particular emphasis on the approach to carbon dioxide (CO_2) emissions from private cars in Europe, followed by a discussion of the implications of tough CO_2 targets for the nature of transport policy and for the structure of the transport sector as such.

The Importance and Role of Transport

As shown in Fig. 3.1, the transport sector accounts for 23 percent of the world's GHG emissions and about 30 percent of the emissions in the developed countries belonging to the Organization of Economic Cooperation and Development (OECD) (IEA 2007, Olivier et al. 2005; 2006). Road transport emissions account for the vast majority of these emissions, or roughly about 75 percent globally and in the OECD. Transport sector emissions have risen strongly between 1990 and 2005, in all regions except many former Eastern Bloc countries. Between 1990 and 2005, transport CO_2 emissions rose by 22.3 percent in the 15 early member nations in the European Union (EU-15), by 44.7 percent in the new EU member states, by 28.7 percent in North America, and by 32.3 percent in OECD nations in Asia.

Projections for the future suggest continued strong growth in transport volumes in all modes, especially in non-OECD countries. As a result, the

J. Short
International Transport Forum, OECD, 2 rue André Pascal, 75775 Paris Cedex 16, France

D. Sperling, J.S. Cannon (eds.), *Reducing Climate Impacts
in the Transportation Sector*, DOI: 10.1007/978-1-4020-6979-6_3,
© Springer Science+Business Media B.V. 2009

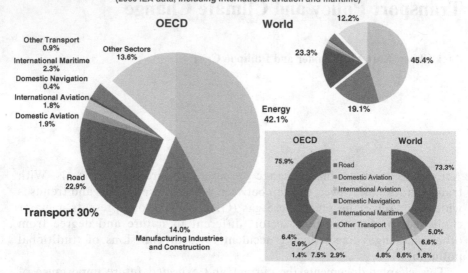

Fig. 3.1 Transport's share of CO2 emissions from fuel combustion
Source: IEA and UNFCCC data.

motorization rate is likely to triple between 2000 and 2050 (WBCSD, 2004), air passenger traffic will be 2.5 times higher in 2025 than in 1985 (Boeing, 2007) and air cargo will be three times higher in 2025 compared to 1995 (Airbus, 2007). Similarly, shipping volumes in million metric tons are on track to increase threefold between 1980 and 2020 (Corbett, 2007). While precise growth rates are uncertain, there is little doubt about continued high-paced growth in transport use.

These figures suggest policy priority should be given to road transport, though the high growth rates in the other modes are also causes for concern. Moreover, cost comparisons with other sectors clearly matter for a balanced abatement strategy as well. Many studies indicate it is more cost effective to reduce emissions from nontransport activities, like domestic heating and cooling or power generation, but it will be politically difficult to justify a do nothing approach in transport, if other sectors are taking measures to limit emissions.

At the same time, it needs to be recognized that the forces driving many of these trends are very powerful, and not easily amenable to steering through policy. This juxtaposition of powerful upward trends for activity and emissions in the sector and political commitments to reduce total emissions by as much as 50 or 60 percent by 2050 poses a formidable challenge. The gap between political expectations for emissions and the trends in the transport sector is widening.

Actions Being Taken

The European Conference of Ministers for Transport (ECMT), now called the International Transport Forum, has done a study entitled "Cutting CO_2 Emissions: What Progress?" that gathered data on actions to reduce CO_2 emissions from almost 50 countries, including most OECD countries as well as countries in central and eastern Europe (ECMT, 2007). The information was obtained mostly from national transport ministries and from databases of the United Nations Framework Convention on Climate Change (UNFRCCC) and the European Union (EU) to which member countries had reported their actions in line with requirements under the Kyoto Protocol. The study identified more than 400 measures that these countries cited in their responses. Table 3.1 categorised them among four groups: demand management, fuel efficiency, carbon intensity and modal split.

Most of the measures fall into the fuel efficiency and modal split categories. For all EU countries, the fuel efficiency category includes a strong emphasis on a voluntary agreement with the automotive industry as well as labelling initiatives, fuel price changes and changes in vehicle taxation. The modal split category includes all the measures cited to support rail or public transport, the traditional policy instruments from transport ministries. A heavy emphasis was also placed on EU biofuel targets and incentives for their introduction. At the time the data were collected in 2006, analysis on biofuels was at an early stage and there was a far more optimistic view about the impacts and cost effectiveness than now (ITF, 2007).

Table 3.1 shows few measures on demand management, suggesting that these strategies apparently have not found their way into mainstream policy making, although they are the staple of climate change conferences. Nevertheless, recent adoption of congestion charging mechanisms in London and Stockholm, and new proposals in New York City and elsewhere suggests there is progress in this area.

Table 3.1. Analysis of over 400 policies

		% of policies
Demand	Urban planning to discourage sprawl; Road pricing; Logistics optimisation.	4
Fuel efficiency – Technical	Tax differentiation to promote EFVs; Vehicle efficiency regulations CAFE, Top-Runner;	31
– On-road	Driver training; Car pooling; Logistics management, route planning/guidance.	16
Carbon intensity	Biofuel targets and tax incentives; Hydrogen fuel cell R&D; Incentives for CNG buses.	24
Modal split	Targeted subsidies for public transport.	28

Table 3.2. Analysis of policies

Top policy combinations	Ave% impact	No. of ITF countries
Fuel taxpolicy	7.1	6
Vehidefuel efficiency/voluntry agreement	4.6	EU+3
Vehide efficiency tax incentives	4.3	17
On road eff.education/training	2.8	11
Biofuelsregulation	2.6	3
Fuel efficiency information	2.2	11
Road pricing	2.1	3

Few robust assessments of the costs or potential CO_2 reductions from the 400 strategies have been completed to date. For example, the rush into biofuels was not supported by proper financial or environmental analysis. This is a major flaw that holds serious consequences. Improvements to the analytical underpinning of climate change policy in transport are clearly desirable and would improve rational policy design. At the same time, it is not clear that the situation is worse in transport than in any other sector.

Table 3.2 shows the abatement potential from different climate change measures. The most cost effective measures to reduce CO_2 are those that charge users for their emissions. Fuel taxes and vehicle registration and excise duties are effective in altering behaviour and in influencing purchasing decisions. In contrast, traditional policy measures, like incentives to switch to public transport, which admittedly have other objectives, are not the most effective in delivering reductions in GHG emissions. While a "co-benefits" approach is a central plank of European transport policy, it is clear that such policies, at the margin, do not deliver enormous CO_2 benefits.

The measures now being implemented or proposed could reduce the total transport CO_2 emissions by about 700 million metric tons compared to what they would have been without the measures. They will not achieve net reduction from current levels, but would reduce future growth by about 50 percent. The analysis is plagued by lack of knowledge of likely costs. It has a possible optimistic bias, as evidenced by the International Energy Agency (IEA) recent upward revision of its transport emissions trend. Countries tend to be optimistic about impacts when reporting to the UNFCCC. On the other hand, progress is being made in countries like France, Germany and Japan, where CO_2 levels have plateaued or are declining in recent years with continued economic growth due to a variety of climate change policies and higher oil prices.

EU Policy Aiming to Reduce Transport CO_2 Emissions

There is now growing concern that CO_2 emissions growth from the transport sector in Europe will not allow it to meet its 2020 emissions target of a 20 percent reduction in GHG levels compared to actual 1990 emissions. This has led the European Commission to act on a number of fronts to reduce the rate of emissions and the overall amount of CO_2 emitted from transport activity.

According to IEA data, EU-27 countries emitted 1,248 million metric tons (Mt) of CO_2 in 2005 from transport activity, including international aviation and maritime transport. This represents a 31.5 percent growth compared to emissions in 1990. If international aviation and maritime emissions are excluded, total EU-27 transport CO_2 emissions grew by 24.5 percent over the same period. The bulk of these emissions are from road transport, yet, as is the case elsewhere, aviation and maritime emissions are a significant and fast growing source. Aviation, including international aviation, represents 12 percent of EU-27 transport CO_2 emissions and has grown by 69 percent since 1990. Maritime, including international maritime, transport represents 14 percent of EU-27 transport CO_2 emissions and has grown by 37 percent since 1990. The European Commission has recently acted to address aviation CO_2 emissions and is reserving the right to address emissions from ocean-going ships should progress on reducing GHG emissions from these prove to be unsatisfactory. This issue is discussed more in the sidebar text box 3.1.

Box 3.1 EU policy on aviation and maritime GHG emissions

While CO_2 emissions from aviation and maritime activity are relatively small, they are growing quickly, prompting the EU to act to curb their emissions. Much of the focus has been on aviation. The European Commission issued a legislative proposal in 2006 to cap these emissions and introduce trading among airlines. Late in 2007, a compromise was reached among EU transport ministers about how this should come about. Key points of the agreement set to come into force in 2012 include:

- Inclusion of all airlines flying to and from the EU. This will most likely trigger legal action by non-EU states.
- Emission levels for each airline will be capped at actual levels between 2004 and 2006. Many EU Parliament members and nongovernment organization representatives had asked for significant cuts going beyond these levels.
- 90 percent of the pollution permits will be distributed for free, raising concerns about windfall profits for airlines for non-existent CO_2 abatement and unfair treatment of new market entrants. On the latter point, however, the current proposal calls for a 3 percent set-aside for new entrants.
- States have authority on the use of the scheme's revenues, but the European Commission proposal suggests that these "should" be spent on GHG mitigation measures.
- The scheme does not address the non CO_2 global warming impacts from aviation emissions of nitrogen oxides and particulates, contrails and cirrus cloud formation.
- Low volume operators from developing countries are exempt from mandatory participation in the scheme.

Box 3.1 (*Continued*)

Maritime GHG emissions are also a concern of the EU's. The European Commission has pledged to interfere should progress on this front at the International Maritime Organization be unsatisfactory, as it has done with the aviation sector. From a policy perspective, the maritime sector is an important one to address since it presents many opportunities for relatively low-cost GHG mitigation measures.

CO_2 Emissions From Cars in Europe: From a Voluntary To a Regulatory Approach

Discussion in Europe on vehicle fuel economy, which began with the energy crises in 1973 and 1979, was largely forgotten as real fuel prices fell in the 1980s, but intensified in the early 1990s following the Brundtland Report on sustainable development and the early evidence on global warming. This concern in the early 1990s did not translate into new transport policies or tougher regulations for vehicle manufacturers. Traditional transport policy instruments, such as investment and subsidies, tended to become subject to more rigorous economic appraisal, though most countries continued the tradition to support rail and public transport, not for reasons of limiting CO_2 emissions, but for a range of broader policy reasons. Voluntary agreements were made between governments and the vehicle manufacturing industry, in the ECMT in 1995 and throughout the EU in 1998. With unprecedented attention being given to climate change, there are many new transport policies governing, for example, future investments in roads or airports. Concrete proposals for legislation on vehicle emissions have been made and are being re-examined by European legislators during 2008.

Voluntary Agreements

In 1995, ECMT ministers agreed with the Organisation Internationale des Constructeurs Automobile (OICA) to make "substantial and continuous cuts in emission of CO_2 from new cars." The text and a monitoring mechanism were agreed at the ECMT Council meeting in Vienna in 1995 (ECMT, 1995).

In 1998, a voluntary agreement was signed between the governments of 15 EU countries and vehicle manufacturers, represented by the major automotive trade associations in Europe and Asia, the ACEA, JAMA and KAMA. This agreement contained a numerical target of 140 grams per kilometre of driving, but there was some flexibility in the implementation dates, which were set for around 2008. The monitoring of the implementation of these agreements was initially carried out by the ECMT and then by the EU. As the implementation period unfolded, the data showed that new car fuel efficiency was improving,

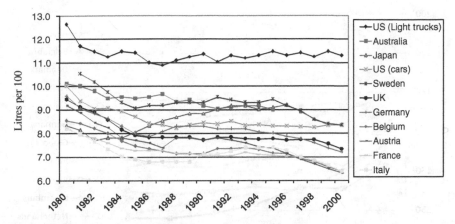

Fig. 3.2 Evolution of new car fuel economy
Source: Lee Schipper, WRI-EMBARQ, private communication.

but not at a pace commensurate with achieving the targets. A principal reason for the improvement has been the massive switch to diesel fuel across Europe.

Figure 3.2 shows the fuel efficiency trends in CO_2 emissions for new cars in several key markets between 1980 and 2000. After a generalized improvement in fuel efficiency in the early 1980s, which itself was the continuation of a trend from the 1970s, regions diverged in their fuel economy performance. Fuel economy performance in the United States and Europe stagnated through the mid 1990s, at which time Europe began to register improved fuel economy once again, while the United States saw no improvement. Japan experienced reduced fuel economy from the mid-1980s to the mid-1990s, at which point resumed its initial path towards improved fuel economy.

More recently, most countries in Europe have seen a continued decrease in the amount of CO_2 emitted per kilometer of car travel. Between 1995 and 2004, there was a decrease of 12.4 percent, from 186 to 163 grams per kilometer in the sales-weighted average emissions of new cars in Europe. However, data from 2005 and 2006 suggest that the improvement has plateaued in many EU countries and that the aggregate 2008 target in the EU agreement will not be met. These trends are shown in Fig. 3.3.

The European Commission and others, including many environmental groups, have been forthright in attributing responsibility for the failure to reach targets to the automotive industry. On the other hand, industry and others have argued that responsibility must also be taken by governments, as they failed to provide the fiscal or other incentives for consumers to purchase more fuel efficient cars. Claims that fuel efficient vehicles, such as the Volkswagen Lupo, were available, but that consumers would not buy them, were suggested as proof of industry's good faith, but there were also accusations that industry had not tried to market these vehicles.

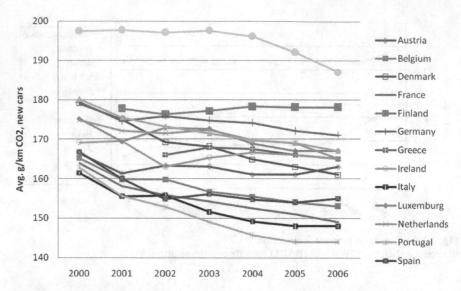

Fig. 3.3 CO2 emissions from new cars in EU-15 countries from 2000-2006
Source: EU and Transport and Environment.

In fact, the targets were unlikely ever to be achieved. The main part of the actual reduction occurred through the switch to diesel. At the same time, trends in purchasing behaviour towards more powerful, bigger, and heavier cars continued in Europe, as well as the United States. Moreover, incomes rose and real fuel prices were declining or stable or at least stayed well below historical peaks up to 2004. These factors were pulling in the direction of less fuel efficient cars, while there were not many specific measures acting in the other direction.

Whatever the value of a voluntary agreement in a non-legislative organization like ECMT, a voluntary agreement in the EU was a clear sign of weakness by legislators. It was not politically possible to pass legislation in the mid-1990s because of the opposition of the industry. As climate change concerns rose and the influence of the European Commission's environmental directorate strengthened, legislative proposals became inevitable.

The environmental directorate in the European Commission was steadfastly opposed to the voluntary agreement. Industry allies in the economic and industrial directorates were strong. Perhaps industry believed that their hegemony would continue, or that attention for climate change would dissipate, it would have been very remiss of industry not to have read the signs. Given the failure of the voluntary agreements and the unprecedented growth in concern for climate change, it was inevitable that legislators would try to strengthen the approach.

While discussion continues between industry and government, the key player remains the consumer. The inability of industry or government to influence consumer vehicle purchase choices has strongly undercut the effectiveness of

GHG reduction policies. Consumers continue to buy more comfort, power and speed, usurping most of the potential fuel economy improvements possible through new technologies.

Consumer tastes and preferences are not beyond influence, however. Governments introduced labelling schemes and gave general encouragement on purchasing fuel efficient cars, but few set the fiscal conditions, such as vehicle excise and registration charges, at levels high enough to strongly influence purchase decisions. For industry, marketing is a major cost component, but studies shows that fuel economy is not a key factor in vehicle choice. Carmakers, therefore, rarely give it attention in their marketing, which rather surprisingly does not yet seem to have changed with high fuel prices. It also appears that the most fuel efficient cars are not the most profitable for industry. Thus, neither consumer preferences nor the bottom line give industry incentives to market these vehicles. Thus, it appears that a combination of government inaction, consumer indifference and industry hostility have conspired to make improving fuel efficiency an uphill struggle.

A New Regulatory Approach

On December 19, 2007, the European Commission published its proposals for legislation on CO_2 emissions for new cars (EU, 2007). Details of this proposal are shown in sidebar text box 3.2. The text is a compromise between the different objectives and views expressed during its preparation. Essentially, a legislative target of 130 grams per kilometer for new vehicles is set for introduction in 2012. A further 10 grams per kilometer reduction are to be obtained from ancillary measures, such as use of low viscosity lubricants, low rolling resistance tires, more efficient air conditioning systems, and greater use of low-carbon fuels. One of the most interesting features of the proposal is the method of dealing with the bigger and heavier cars, which at present have emissions well above the target level of 130 grams per kilometer.

Box 3.2 Overview of the 2007 EU fuel economy regulation proposal

In December 2007, the European Commission presented its proposal for regulating passenger car fuel economy as of 2012. The proposal's main characteristics are as follows:

- Average emissions from the new passenger car fleet in the EU are limited to 130 grams of CO_2 per kilometer as of 2012.
- An additional 10 grams of CO_2 per kilometer is to be obtained from complementary measures.
- A passenger car's specific standard depends on its weight, as heavier cars are allowed to emit more. The car specific standard increases linearly with weight, as depicted in the Figure 3.4. Comparison with the 2006

Box 3.2 (*Continued*)

relation between weight and emissions shows that the standard is tighter for heavier cars.

- The choice for weight over footprint is motivated by the observations that weight correlates well with current emissions, and data on weight are readily available. There is a provision calling for collection of data on footprint, such as track width by wheelbase, suggesting future use of that criterion.
- There is a provision that the formula for the standard is adapted to potential weight increases between 2006 and 2009, in order to discourage manipulation of weight. Specifically, the European Commission proposes to track weight changes from 2006 to 2009, and to extrapolate any trend of weight increases to 2012. It will adapt the formula to make sure the target of 130 grams of CO_2 per kilometer is attained for the new weight base.
- A manufacturer's specific standard is the weighted average of the standards of the new cars it sells in a given year.
- Manufacturers are allowed to form agreements to jointly attain the standard. These agreements are limited in time to a five-year maximum, but are renewable, and participants cannot exchange any other information than that required to calculate standards and fines. This mechanism provides additional flexibility, and reflects the policy goal of attaining fleetwide average reductions to 130 grams of CO_2 per kilometer by 2012.
- Manufacturers' non-compliance leads to fines, which are calculated as follow:
  - Excess emissions are the difference between a manufacturer's average emissions on its sales in a given year, minus the manufacturer's specific standard for that year;
  - Penalties are defined per gram of excess emissions, and increase over time from 20€ per gram in 2012, to 35€ in 2013, to 60€ in 2014, and to 95€ in 2015;
  - The fine per average car is multiplied by the manufacturer's sales.
- Fine revenues accrue to the general EU budget.

Some examples may clarify the stringency of the fines. Supposing that there is no weight increase between 2006 and 2009, the formula for calculating the car-specific standard (CST, in grams of CO_2 per km) is: CST = $130 + 0.0457(M-1289)$, where M is the car's weight. Supposing in addition that manufacturers produce the same sales mix with the same fuel economies as in 2006 (this obviously is not the idea), we can calculate the average fine per vehicle sold. A producer whose sales mix reflects the 2006 EU average of weight (1,289kg) and CO_2 emissions (about 160g of CO_2 per km) would not meet the target of 130g of CO_2 per km, resulting in a fine of 600€ per car in 2012, rising to 2,850€ per car in 2015. Another example shows that producing lighter but less fuel efficient cars is strongly discouraged: with an average weight of 1,250kg and CO_2 emissions of 170 g per km, the average fine per car is 835€ in 2012 and 3,967€ in 2015.

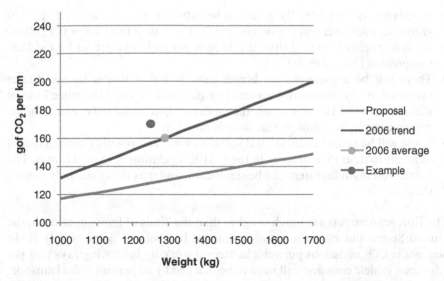

Fig. 3.4 Proposed EU CO_2 regulation for passenger cars and 2006 weight-emission ratio

The approach is illustrated schematically in Fig. 3.4. In essence, all vehicle types are to reduce emissions, but relatively greater reductions are to come from bigger and heavier cars. This is shown by the reduced slope of the lower line, which plots allowed emissions against weight, compared to 2006 market conditions, shown in the upper line.

Adjustment of CO_2 limits by vehicle weight is the subject of debate, with environmental groups and parts of the vehicle industry casting it as a victory for manufacturers of large cars, because heavier cars are allowed to emit more. A one-size-fits-all 130 gram limit applied to all new vehicles would have been completely unrealistic for some manufacturers, however, and was certainly not politically feasible. The approach is reminiscent of that adopted in the U.S. corporate average fuel economy (CAFE) regulation for light trucks, where fuel economy was defined for vehicle classes on the basis of their footprint. In the CAFE program, footprint was chosen over weight as it is harder to change for manufacturers, thereby reducing the scope for strategic manoeuvrings by manufacturers.

To give the legislation teeth, a system of fines has been proposed. These fines, described in the sidebar, are, at least at first sight, rather high. Clearly the intention is to dissuade industry from opting to pay the fines rather than meeting the targets. The economic foundation for these fines is not entirely clear, however, as this depends on how production cost increases compared to the payment of fines. The payment of fines into the general EU budget will undoubtedly be a controversial topic as well.

The proposals are under debate by the European Parliament and Council. Speculation on the outcomes is difficult, but a number of points can be made:

- Regulation is undoubtedly going to be introduced, at or around the 130 grams per kilometer level, which is roughly equivalent to 45 miles per gallon, a much tougher standard than the 35 mile per gallon target set by the U.S. Congress in December 2007.
- There will be allowances for bigger cars. Whether this is by weight, as proposed, or by footprint, measured by the trackwidth of the wheelbase, is still not certain. There is a risk that using weight alone may lead to some perverse incentives, though this issue is not clear.
- There will be some beneficial and some perverse and possibly unanticipated elements to it, as there were with the CAFE regulation in United States. The impacts on manufacturers can be significant, and may differ strongly among them.

The European targets are much tougher than the kinds of limits in effect in the United States, but even the trend in Europe has been to achieve only slight declines in CO_2 emissions per vehicle. Recent work by Julia King says that per kilometer vehicle emission will need to be reduced by 90 percent if the transport sector is to contribute substantially to meeting CO_2 emissions targets needed to prevent significant climate change (King, 2007). This report argues that cost effective reductions of up to 50 percent per vehicle are possible in the short term. From this perspective the targets are not too tough. Similarly, Steve Plotkin also argues that up to 50 percent reduction in vehicle emissions are possible and cost effective for the U.S. fleet (Plotkin, 2007).

The pooling mechanism provides potential flexibility. Over-attainment becomes potentially valuable, possibly reducing the incentives for high-performing manufacturers to reduce fuel economy. In addition, low-performing manufacturers now face a choice of either paying fines, pooling with high-performing firms and possibly making side-payments to them, or improving fuel economy.

Non-compliance is punished, but there is no mechanism to reward manufacturers that go beyond the target of 130 grams of CO_2 per kilometer target. For example, there is no direct financial incentive for over-compliance, although over-compliant manufactures may benefit through the pooling provision. The European regulation also does not contain a mechanism for automatic tightening of the standard.

The new EU standards will make a useful contribution to meeting targets for climate stabilization, but they are only the beginning. Far sharper reductions will be needed over the decades ahead if trends are to be broken and targets for reductions met. The consequences for the vehicle industry can only be guessed. Industry rationalization and consolidation, occurring in any case, are likely to change only in some details. Italian and French car makers are likely to adapt more easily than competitors in Germany or Sweden. With Germany playing a leading role in the climate discussions in the EU and among the G8 group of economic leaders, opposition from the car industry in Germany was originally muted. But the German industry has more recently identified the risks and has lobbied intensively, resulting in the weight differentiation proposals.

Consequences for Transport Policy

At a general level, the challenge to reduce emissions by about 50 percent over the next half century in the face of projected growth of 50 percent is enormous. Some believe the solution is essentially technological. Given the gap between political aspirations and ongoing trends, technology certainly needs to play a central role. In this regard, some analysts suggest that regulating fuel economy with standards provides no incentive for the adoption of alternative technologies unless the standards are very strict, so that complementary fiscal measures are justified (e.g. Small and Van Dender, 2007). For others, including many European countries, technology will be important, but not sufficient, as transport policy and especially demand management measures are needed in strong supplementary roles.

It is ironic that CO_2 reductions that have been achieved to date have not been the least expensive to accomplish. Instead, climate change policies reward more expensive and less cost effective measures. At present, neither technology nor policy can cheaply deliver the kind of reductions being sought. Given the underlying growth pressures from mobility, the reductions in emissions per kilometer will need to far exceed the current 50 percent target. It is not yet known how to do this.

There is a clear need for improved analytics. Many countries claim credit for their traditional measures without knowing either the costs or the impacts needed to calculate the price of carbon reductions from different strategies. Transport project analysis already uses such explicit pricing mechanisms, with valuations for a life set by examples based on worker safety, for example. A price for carbon, perhaps set at 50 euros a metric ton, will allow a more rational approach to the topic and will help avoid the lurches in policy that are common as priorities change. Making the price of carbon explicit also has the additional benefit of reducing uncertainty over business costs, and this will help industry to make sound decisions on major investment regarding less carbon-intensive fuels.

Beyond the analytics is the issue of the institutional framework for climate change policies. Transport is a broad area and the demands for it are determined by many factors. Transport policy in most countries is concerned primarily with issues of investment in infrastructure and regulation of the operators and companies in the sector. There is a wide array of agencies and ministries involved in tackling climate change from transport. Transport ministries or their agencies are usually not responsible for technical standards relating to air pollution or for many of the technical standards related to vehicle standard conformity. They are generally not responsible for land use or fiscal decisions. Transport ministries need to be at the center of climate change debates and play a proactive role there. Furthermore, they need to strengthen their analytic abilities and increase their coordination role.

There are many measures that can be taken to improve the efficient organization of transport activities that will reduce CO_2 emissions. Attempting to transform transport for climate's sake is not likely to work and is certainly extremely costly. Transport policy should not become solely energy policy.

References

Airbus (2007), Global Market Forecast: The Future of Flying 2006–2025, Airbus Industries, Toulouse, France

Boeing (2007), Current Market Outlook – 2007, Boeing Corporation, Seattle, Washington

Corbett, James, 2007, Scoping Paper on Emissions from the Maritime Shipping Sector, prepared for the OECD-International Transport Forum Joint Transport Research Centre Working Group on GHG Reduction Strategies from the Transport Sector, 21–22 May, 2007, Paris, France

European Conference of Ministers for Transport (ECMT) Council meeting in Vienna, Austria, in 1995

European Conference of Ministers for Transport (ECMT), 2007, Cutting Transport CO2 Emissions: What Progress?, European Conference of Ministers of Transport, Paris, France

European Union (EU), 2007, Proposal for a Regulation of the European Parliament and of the Council: Setting emission standards for new passenger cars as part of the Community's integrated approach to reduce CO2 emissions from light duty vehicles, COM(2007) 856 Final

International Energy Agency (IEA), 2007, CO_2 Emissions from Fuel Combustion 1971–2005, 2007 Edition, International Energy Agency (IEA), Paris, France

International Transport Forum (ITF), http://www.internationaltransportforum.org/jtrc/RTbiofuelsSummary.pdf, website accessed in 2007

King, Julia, 2007, The King Review of Low Carbon Cars: Part I – The Potential for CO_2 Reduction, UK HM Treasury, London

Olivier, Jos G.J., John A. Van Aardenne, Frank Dentener, Valerio Pagliari, Laurens N. Ganzeveld, and Jeroen A.H.W. Peters, 2005, Recent trends in global greenhouse gas emissions: regional trends 1970–2000 and spatial distribution of key sources in 2000. Environmental Science, 2(2–3), pp. 81–99

Olivier, Jos G.J., Tinus Pulles, John A. Van Aardenne, 2006, Part III: Greenhouse gas emissions: 1. Shares and trends in greenhouse gas emissions; 2. Sources and Methods; Greenhouse gas emissions for 1990, 1995 and 2000. In CO2 emissions from fuel combustion 1971–2004, 2006 Edition, pp. III.1–III.41. International Energy Agency (IEA), Paris, France

Plotkin, Steve. 2007. A paper submitted to the International Forum, accessed at http://www.internationaltransportforum.org/jtrc/DiscussionPapers/DiscussionPaper1.pdf

Small and Van Dender, 2007, accessed at http://www.internationaltransportforum.org/jtrc/DiscussionPapers/DiscussionPaper16.pdf

World Business Council for Sustainable Development (WBCSD), 2004, Mobility 2030: Meeting the Challenges to Sustainability, World Business Council for Sustainable Development, Geneva, Switzerland

Chapter 4
Factor of Two: Halving the Fuel Consumption of New U.S. Automobiles by 2035

Lynette Cheah, Christopher Evans, Anup Bandivadekar and John Heywood

The United States (U.S.) transportation sector is almost totally dependent on gasoline and diesel fuel refined from oil to provide the remarkable mobility provided by the automobile. This dependence presents a challenging energy and environmental problem, as the transportation sector is responsible for two-thirds of total petroleum consumption and a third of the nation's carbon emissions. Amid growing concerns over energy security, and the impacts of global climate change, the U.S. Congress in December 2007 adopted a new legislative directive to increase the fuel economy of new passenger vehicles over the next two decades.

Transportation experts at the Laboratory for Energy and Environment, part of the Massachusetts Institute of Technology (MIT) in Cambridge, Massachusetts, have examined the vehicle design and sales mix changes necessary to double the average fuel economy or halve the fuel consumption of new light duty vehicles (LDVs)—including cars, wagons, sport utility vehicles (SUVs), pickup trucks and vans—by model year 2035. To achieve this factor-of-two target, three technology options that are available and can be implemented on a large scale were evaluated:

- Channeling future vehicle technical efficiency improvements to reducing fuel consumption rather than improving vehicle performance
- Increasing the market share of diesel, turbocharged gasoline and hybrid electric gasoline propulsion systems
- Reducing vehicle weight and size.

The scenarios developed and analyzed at MIT demonstrate the challenges of this factor-of-two improvement. They reveal that major changes in all three options need to be implemented before the target is met. A steady rate of progress toward the target would reduce the fuel used by LDVs by roughly one-third in the year 2035. The sales-weighted average fuel economy calculated

L. Cheah
Sloan Automotive Laboratory, 31-153, Massachusetts Institute of Technology,
77 Massachusetts Avenue, Cambridge, MA 02139-4307, USA

D. Sperling, J.S. Cannon (eds.), *Reducing Climate Impacts
in the Transportation Sector*, DOI: 10.1007/978-1-4020-6979-6_4,
© Springer Science+Business Media B.V. 2009

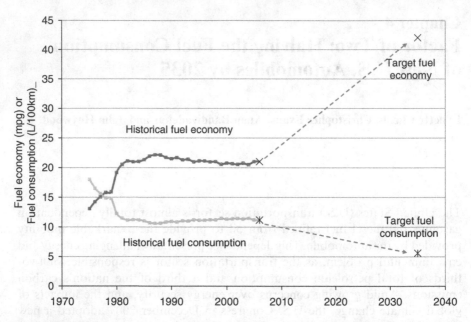

Fig. 4.1 Sales-weighted average new vehicle fuel economy (FE) and fuel consumption (FC)

by the U.S. Environmental Protection Agency (EPA) would increase from 21 miles per gallon (mpg) in 2007 to 42 mpg by 2035 as shown in Fig. 4.1. Adjusted, combined 55/45 percent city/highway EPA laboratory test fuel economy and fuel consumption numbers were used throughout the MIT analysis. In terms of fuel consumption, this is equivalent to halving the average amount of fuel vehicles consume to travel a given distance, or reducing today's 11.2 liters per 100 kilometers of driving (L/100 km) fuel consumption to 5.6 L/100 km by 2035.

To achieve this target, the analysis evaluated combinations of available fuel-saving technologies and then considered their associated increased costs of production. The impact on greenhouse gas (GHG) emissions, in particular carbon dioxide (CO_2), on a lifecycle basis was also evaluated. By illustrating scenarios of how fuel consumption reductions can be attained in automobiles, this study provides a useful reference for both policymakers and the automotive industry.

Background and Approach

Roughly 16 million new vehicles are introduced onto the roads in the U.S. each year. Almost half of new vehicles sold are passenger cars, while the others are light trucks. More than 95 percent of vehicles operate on gasoline, using conventional, naturally-aspirated, spark-ignited internal combustion engines.

Today, the average new car consumes 9.6 L/100 km of gasoline, equivalent to a fuel economy of 25 mpg, and can accelerate from 0 to 100 km per hour, or 0 to 60 miles per hour (mph), in under 10 seconds. The average car weighs 1,620 kilograms (kg), or 3,560 pounds, mostly in iron and steel, and offers 3.25 cubic meters, or 114 cubic feet, of interior room for both the passengers and their cargo. The average light truck weighs 2,140 kg, or 4,720 pounds, and consumes 12.8 L/100 km, or 18 mpg.

One approach to improve vehicle fuel efficiency is to improve conventional vehicle technology. For example, gasoline direct injection, variable valve lift and timing, and cylinder deactivation can individually realize efficiency improvements by 3–10 percent, and are already being deployed in gasoline spark-ignition engines. Further efficiency improvements from dual clutch and continuously variable transmissions are likely to occur in the near future, as well as reductions in aerodynamic drag, and rolling resistance (Kasseris and Heywood, 2007).

Another approach is to use alternative powertrains, such as turbocharged gasoline engines, high speed turbocharged diesel engines, and hybrid-electric systems. These alternatives provide additional fuel efficiency over naturally-aspirated gasoline engines. A turbocharger allows an engine to be downsized by increasing the amount of air flow into the engine cylinders, while delivering the same power. Diesel engines operate by auto-igniting diesel fuel injected directly into a cylinder of heated, pressurized air. This allows a high compression ratio, enables combustion with excess air, and eliminates throttling losses to offer increased engine efficiency. A hybrid-electric drivetrain provides the ability to store energy in a battery and run using power from both an engine and electric motor. This offers improved efficiency by decoupling the engine from the drivetrain at lighter loads where the efficiency is low, by turning the engine off while idling, and by storing and reusing much of the vehicle's kinetic energy captured during regenerative braking. These attributes also allow secondary benefits from downsizing to a smaller, lighter engine (Kromer and Heywood, 2007).

Figure 4.2 shows the current and future fuel consumption benefit of using these alternative drivetrains in the average passenger car and light truck, with today's naturally-aspirated gasoline internal combustion engine as the reference. The best-selling car and light truck, the Toyota Camry CE mid-size sedan and Ford F150 pickup truck, were selected as representative models for this analysis. The hybrid electric vehicle (HEV) model assessed was a full hybrid with a parallel architecture, which for cars, is similar to a Toyota Camry Hybrid. It offers the highest potential fuel savings, although the robust performance of diesels over a variety of operating conditions may make them more suitable than hybrids in heavy towing applications. Full HEV systems have more powerful electric drives that assist the engine, and allow limited driving without use of the engine.

Over the next three decades, if all improvements to conventional vehicle technology are focused on reducing fuel consumption, significant benefit can be

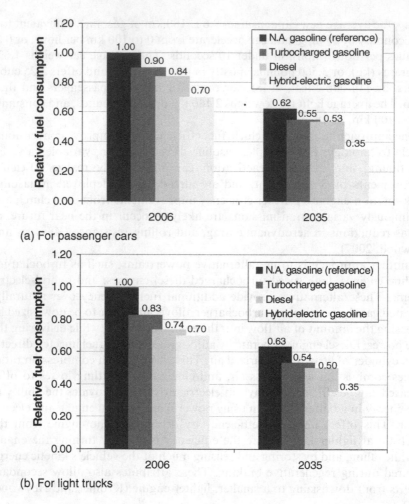

(a) For passenger cars

(b) For light trucks

Fig. 4.2 Current and future relative fuel consumption of alternative powertrains (Kasseris and Heywood 2007, Kromer and Heywood 2007)

realized across all powertrain options, including vehicles that continue to use the conventional naturally-aspirated gasoline engine.

Next, vehicle weight reduction can reduce the overall energy required to accelerate to a given speed. Reductions in weight can be achieved by a combination of material substitution, vehicle redesign, and vehicle downsizing. Material substitution involves replacing heavier iron and steel with weight-saving materials like aluminum, magnesium, high-strength steel, and plastics and polymer composites. Redesign reduces the size of the engine and other components as vehicle weight decreases, or through packaging improvements, which reduce exterior vehicle dimensions, while maintaining the same passenger and cargo

Fig. 4.3 The vehicle design and marketing options to reduce fuel consumption

space. Downsizing can provide further weight reduction by shifting sales away from larger and heavier to smaller and lighter vehicle categories.

When considering various ways of achieving the target of halving fuel consumption in vehicles, the MIT researchers chose to focus on options that are on the commercial market today, and which do not require significant changes to the fueling infrastructure. For this reason, plug-in hybrid electric, battery electric or hydrogen fuel cell vehicles were not considered, although they are potentially important technologies for realizing vehicle fuel consumption reductions. Fuel alternatives were also deliberately excluded, although some alternative fuels can offer reductions in petroleum use and GHG emissions. Thus, three options, shown in Fig. 4.3 and discussed in more detail below, were explored based on their current feasibility, availability, and market-readiness:

- Emphasis on reducing fuel consumption—dedicating future vehicle efficiency improvements to reducing fuel consumption, as opposed to improving vehicle performance
- Use of alternative powertrains—increasing market penetration of more efficient turbocharged gasoline engines, diesel engines, and hybrid electric-gasoline drives
- Vehicle weight and size reduction—additional weight and size reduction for further fuel efficiency gains

Option #1: Emphasize Reducing Fuel Consumption

The first option is to emphasize reducing fuel consumption over improving the vehicle's horsepower and acceleration, while assuming that vehicle size remains constant. This is an explicit design decision to dedicate future advances in vehicle efficiency into reducing fuel consumption, rather than improving performance. Over the past two decades, more emphasis has been placed on the latter, while the average new vehicle's fuel consumption has remained almost stagnant. If the performance trend of the past two decades continues, the average new car in 2035 could potentially boast 320 horsepower and a 0-to-60 mph acceleration time of 6.2 seconds, outperforming today's BMW Z4 Roadster sports car.

It is questionable whether this level of performance is necessary, or even safe for the average driver on regular roads, regardless of whether the future consumer truly wants or expects this. Speed and horsepower have always had strong marketing appeal, however, and demand might well continue. A trade-off between increasing performance, size, and weight versus reducing fuel consumption must be made in future vehicles. While holding size constant, the MIT researchers defined this trade-off as the degree of emphasis on reducing fuel consumption (ERFC), where:

$$\% ERFC = \frac{\text{Future fuel consumption reduction realized}}{\text{Future fuel consumption reduction possible with constant size and performance}}$$

At 100 percent ERFC, all of the steady improvements in conventional technology over time are assumed to realize reduced fuel consumption, while vehicle performance remains constant. This includes an assumption that vehicle weight will reduce by 20 percent. In contrast, without any emphasis on reducing fuel consumption, in other words 0 percent ERFC, the fuel consumption of new vehicles will remain at today's values, no weight reduction will occur, and all of the efficiency gains from steady technology improvements will be channeled to improve the horsepower and acceleration performance.

By simulating the future vehicles described, using the ADVISOR computer software marketed by AVL, the current and future new vehicle characteristics at different levels of ERFC are obtained and summarized in Table 4.1. These numbers are assessed for spark-ignited, naturally-aspirated gasoline vehicles with an internal combustion engine. The data for alternative powertrains will be different. The trade-off between acceleration performance and fuel consumption for the average car and light truck of a fixed size is depicted in Fig. 4.4.

When full emphasis is placed on reducing fuel consumption (100 percent ERFC). the fuel consumption of a future new car declines by 35 percent from today's value, from 9.6 to 6.0 L/100 km. About a quarter of this fuel consumption reduction is accredited to the 20 percent reduction in vehicle weight. This weight assumption is based on what is feasible in 2035, given the priority placed on achieving lower fuel consumption. If only half of the efficiency gains are used to emphasize lowering fuel consumption, or at 50 percent ERFC, then only half of the total plausible reduction in fuel consumption will be realized by 2035. Future vehicle curb weight is assumed to scale linearly with percent ERFC, so vehicle weight at 50 percent ERFC reduces by 10 percent from today.

Table 4.1 Summary of current and future naturally-aspirated gasoline vehicle characteristics

Year	% ERFC	Fuel consumption (L/ 100 km) [relative]	Horsepower [relative]	0–60 mph acceleration time (s)	Vehicle weight (kg) [relative]
2006	–	9.6 [1.00]	198 [1.00]	9.5	1,616 [1.00]
2035	0%	9.6 [1.00]	324 [1.64]	6.2	1,616 [1.00]
	50%	7.8 [0.81]	239 [1.21]	7.2	1,454 [0.90]
	100%	6.0 [0.62]	151 [0.76]	9.5	1,293 [0.80]

(a) For cars

Year	% ERFC	Fuel consumption (L/100 km) [relative]	Horsepower [relative]	0–60 mph acceleration time (s)	Vehicle weight (kg) [relative]
2006	–	12.8 [1.00]	239 [1.00]	9.9	2,137 [1.00]
2035	0%	12.8 [1.00]	357 [1.49]	7.1	2,137 [1.00]
	50%	10.4 [0.81]	275 [1.15]	8.1	1,923 [0.90]
	100%	8.1 [0.63]	191 [0.80]	9.8	1,710 [0.80]

(b) For light trucks

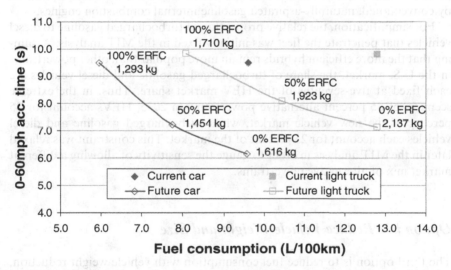

Fig. 4.4 Trade-off between acceleration time and fuel consumption in average new vehicles in 2035. Current vehicle characteristics plotted for reference

Option #2: Use Alternative, More Efficient Powertrains

Today, less than 5 percent of the new vehicles in the U.S. market are turbo-charged gasoline, diesels, or hybrid electrics, but their market shares are expected to grow. Increasing the market penetration of these alternative powertrains, especially the more efficient HEVs, can help achieve a factor-of-two reduction in fuel consumption. The overall benefit obtained from alternative powertrains depends upon how quickly these new technologies can penetrate the existing vehicle fleet.

HEV sales in the U.S. have already grown from 6,000 in 2000, the first full year of hybrid electric vehicle sales, to 213,000 in 2006 (Heavenrich 2006). More diesel passenger vehicle models are expected to be made available in the U.S. from 2008. In Europe, the share of diesel cars grew at an average rate of 9 percent per year from 1990 to 2006 to capture about half of the market today, motivated by innovations in common rail injection and lower taxation of diesel fuel over gasoline. Other automotive technologies such as front or 4-wheel drive and automatic transmission have diffused into the U.S. market at a rate of 7–11 percent per year over 15–20 year periods in the recent past.

Based on these observations, the MIT researchers assumed that the maximum compounded annual growth rate of alternative powertrains in the U.S. market will be 10 percent per year. This corresponds to a maximum 85 percent share of alternative powertrains in new vehicle sales in 2035. If turbocharged gasoline engines, diesels and hybrids are aggressively promoted, for example, only 15 percent of new vehicles introduced onto the roads in 2035 will remain powered by conventional, naturally-aspirated gasoline internal combustion engines.

For simplification, the relative proportion of turbocharged gasoline to diesel vehicles that penetrate the fleet was initially fixed in the MIT analysis. Assuming that the more efficient hybrids remain more popular than other powertrains in the U.S. market, the share of turbocharged gasoline and diesel vehicles are each fixed at five-sevenths of the HEV market share. Thus, in the extreme scenario of 85 percent alternative powertrains in 2035, HEVs account for 35 percent of the new vehicle market, while turbocharged gasoline and diesel vehicles each account for 25 percent of the market. This constraint was relaxed later in the MIT analysis in order to gauge the sensitivity of allowing a different market mix of alternative powertrains.

Option #3: Reduce Vehicle Weight and Size

The third option is to reduce fuel consumption with vehicle weight reduction, beyond what has been assumed at different levels of ERFC. Of the lightweight material candidates available for material substitution, aluminum and high-strength steel (HSS) are more cost-effective at large production volume scales, and the MIT researchers believe their increasing use in vehicles is likely to continue. Cast aluminum is best suited to replace cast iron components, stamped aluminum for stamped steel body panels, and HSS for structural steel parts. Plastics and polymer composites are also expected to replace some steel in vehicles, but to a smaller degree given the higher costs of these materials. With aggressive use of these substitute materials, up to 20 percent reduction in

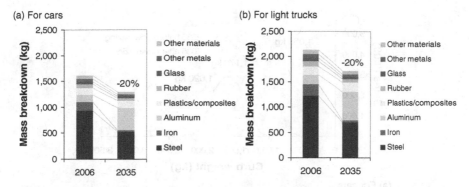

Fig. 4.5 Material composition of the average new gasoline vehicle after material substitution

vehicle weight can be achieved, and the corresponding material breakdown of the average new future vehicle is shown in Fig. 4.5 and Table 4.2.

Redesigning the vehicle includes optimal sizing of subsystems that depend on total vehicle weight. As vehicle weight decreases, the performance requirements of the engine, suspension, and brake subsystems are lowered and they can be downsized. Vehicle redesign may also include "creative packaging" or downsizing the exterior dimensions, while maintaining the same interior passenger and cargo space. The MIT analysis assumed that the weight savings obtained from vehicle redesign are half of that achieved by material substitution.

Beyond material substitution and vehicle redesign, the analysis assumed that an additional 10 percent reduction in the sales-weighted average new vehicle weight is possible through vehicle downsizing. The current difference in weight achieved from downsizing a car by one U.S. EPA size-class ranges from 8 to 11 percent. This can be achieved, for example, by downsizing from a midsize car, like the Toyota Camry, which has between 110 and 120 cubic feet interior volume, to a small car, like the Toyota Corolla, with less than 110 cubic feet interior volume. Only heavier vehicle classes were targeted for downsizing in the MIT analysis, however, while the smaller and lighter vehicles are were not downsized.

Table 4.2 Material composition of the average new gasoline vehicle after material substitution

Material	Cars		Light trucks	
	In 2006, kg	In 2035, kg	In 2006, kg	In 2035, kg
Steel	929	670	1,228	885
Iron	168	82	222	108
Aluminum	142	323	188	427
Rubber	76	61	101	80
Plastics/composites	131	137	173	181
Glass	50	40	67	53
Other metals	55	44	73	58
Other materials	65	52	86	69
Total	1,616	1,408	2,137	1,862

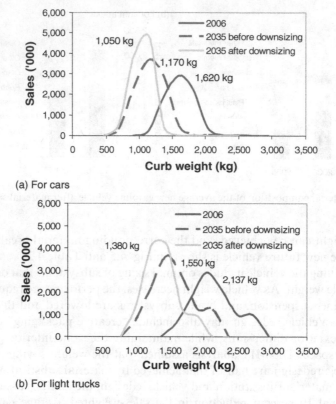

(a) For cars

(b) For light trucks

Fig. 4.6 Current and future new vehicle sales distribution, before and after vehicle downsizing. Average new vehicle curb weight denoted in kilograms

Figure 4.6 shows the sales distribution of new cars today and in year 2035. After material substitution and vehicle redesign without downsizing, the entire future car sales distribution shifts to the lighter weight ranges with no change in its shape. With downsizing, smaller and lighter vehicles will dominate the marketplace, resulting in a lower average weight. The share of light trucks in the 2035 new vehicle fleet was assumed to remain at today's value of 55 percent.

Based on these assessments of aggressive material substitution, vehicle redesign, and downsizing, a maximum weight reduction of 35 percent is possible by 2035. Given the need and demand for weight-adding safety features and passenger and cabin space, it is unlikely that average vehicle weight will decline beyond this. Thus, the minimum average new car weight in 2035 is projected to be 1,050 kg, or about one metric ton, down from 1,620 kg today. The minimum average new light truck weight will be 1,390 kg, a reduction of 750 kg from today's average of 2,140 kg.

Using AVL ADVISOR computer simulations of representative vehicles, the MIT researchers estimated the fuel consumption benefit provided by a given

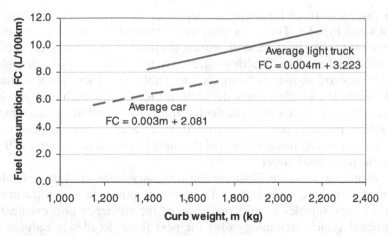

Fig. 4.7 Weight-fuel consumption relationship for future vehicles

reduction in vehicle curb weight. For every 100 kg weight reduction, the adjusted fuel consumption was found to decrease by 0.3 L/100 km for cars, and by 0.4 L/ 100 km for light trucks, as shown in Fig. 4.7. In other words, for every 10 percent weight reduction, the vehicle's fuel consumption reduces by 6 to 7 percent.

Scenario Results

The MIT researchers discovered that exercising each of the three options individually was not sufficient to achieve a doubling of the average fuel economy of new light duty vehicles. Table 4.3 expresses the effectiveness of each option in reducing fuel consumption, if each is exercised independently to its limit. None of them will result in the desired 50 percent fuel consumption reduction on their own. In order to halve the fuel consumption of new vehicles by 2035, scenarios which combine the effects of these options must be developed.

Table 4.3 The effectiveness of the 3 technical options in reducing fuel consumption

Option	Limit	Resulting fuel consumption reduction at the limit (%)
(1) Degree of emphasis on reducing fuel consumption (ERFC)	100% ERFC	36
(2) Increase use of alternative powertrains	Captures up to 85% of the market	23
(3) Vehicle weight reduction	Up to 35% total vehicle weight reduction	19

The results of three bounding, or limiting, scenarios are summarized in Table 4.4 and Fig. 4.8. These scenarios were obtained by exercising two of the three options to their limits, and then using the third option, if needed, until the target was reached. The resulting effects on the 2035 average new vehicle characteristics are shown as "outputs," in Table 4.4. These three scenarios bound the shaded solution space depicted in Fig. 4.8, for both cars and light trucks. Scenarios that lie within the shaded area, which combine greater emphasis on vehicle performance, less weight reduction, and less market penetration of alternative powertrains than each of the three bounding conditions, will also achieve the prescribed target.

The bounding scenarios illustrate the necessary trade-off between vehicle performance, weight, and degree of alternative powertrain penetration. In Scenario I, new vehicles in 2035 realize all of the efficiency improvements in conventional vehicle technology over the next three decades in reduced fuel consumption. They have the same acceleration as vehicles today. On average, vehicles in this scenario weigh one-third less than today, through a combination of aggressive material substitution, redesign, and a 10 percent reduction in size. One out of every three new vehicles sold are propelled by alternative powertrains, while the remaining are powered by naturally-aspirated gasoline engines. Ten percent are turbocharged gasoline, 10 percent are diesel, and 14 percent are HEVs.

In Scenario II, alternative powertrains penetrate much more aggressively into the fleet, achieving an 85 percent market share of new vehicle sales in 2035. HEVs account for 35 percent of new vehicle sales, while diesel and turbocharged gasoline powertrains account for one-quarter each. Only 15 percent of new vehicles sales are comprised of conventional naturally-aspirated gasoline vehicles. Almost all of the conventional technology improvements remain directed towards reducing fuel consumption, and the average weight of new vehicles is reduced by roughly 20 percent.

Finally, Scenario III describes a 2035 sales mix where a moderate level of emphasis is placed on reducing fuel consumption through improvements in vehicle technology. About 60 percent of these improvements are directed towards faster acceleration, lowering the new car average zero to 100 km per hour acceleration time from 9.5 to 7.6 seconds. In order to meet the fuel consumption target, this scenario requires aggressive penetration of alternative powertrain vehicles and maximum weight reduction. Only 15 percent of new vehicle sales are conventional naturally-aspirated gasoline vehicles; 35 percent are HEVs, and the remaining 50 percent is split evenly between turbocharged gasoline and diesel vehicles. Similar to Scenario I, the average vehicle weight is one-third less than today's average in 2035 as a result of aggressive material substitution, vehicle redesign, and downsizing.

These three bounding scenarios reveal trade-offs necessary to halve the fuel consumption of all new vehicles within the constraints of this assessment:

Table 4.4 Results – Scenarios that halve the fuel consumption of new vehicles in 2035

Scenarios	INPUTS Degree of each option						OUTPUTS (vehicle characteristics)					
	% ERFC	2035 powertrain mix				% total. weight reduction from today	2035 average new car			2035 average new light truck		
		Gas natural. Aspirated (%)	Gas turbo (%)	Diesel (%)	Hybrid (%)		0–60mph acc. time	FC, L/ 100 km	Vehicle weight	0–60mph acc. time	FC, L/ 100 km	Vehicle weight
2006 values	–	95	1	2	2	–	9.5s	9.6	1,616 kg	9.9s	12.8	2,137 kg
I. Strong emphasis on reducing FC and vehicle weight	100	66	10	10	14	35	9.4s	4.8	1,054 kg	9.8s	6.4	1,394 kg
II. Strong emphasis on reducing FC and aggressive penetration of alternative powertrains	96	15	25	25	35	19	9.2s	4.8	1,318 kg	9.6s	6.4	1,743 kg
III. Aggressive weight reduction and penetration of alternative powertrains	61	15	25	25	35	35	7.6s	4.9	1,060 kg	8.4s	6.3	1,402 kg
IV. Scenario with aggressive hybrid penetration	75	15	15	15	55	20	8.1s	4.8	1,302 kg	8.8s	6.3	1,722 kg

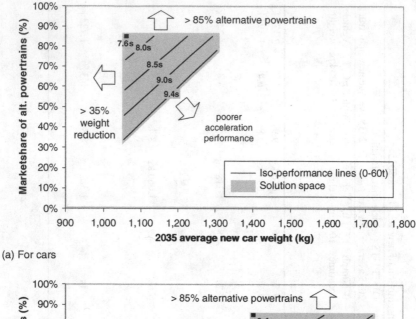

(a) For cars

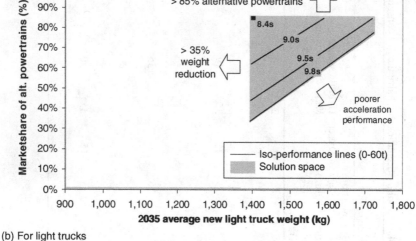

(b) For light trucks

Fig. 4.8 Results – Solution space for Scenarios I, II and III

- The factor-of-two target can be met with lower levels of market penetration of alternative powertrains, but only with full emphasis on reducing fuel consumption and maximum possible weight reduction, including some downsizing (Scenario I).
- To realize a factor-of-two reduction in fuel consumption with a moderate amount of weight reduction and no downsizing, alternative propulsion systems must penetrate the marketplace at a high rate while maintaining today's vehicle performance (Scenario II).

- If performance of vehicles is to be improved significantly above today's level, maximum market penetration of alternative propulsion systems and a large degree of weight reduction and downsizing needs to be achieved (Scenario III).

To illustrate the effects of an alternative powertrain mix, a fourth scenario was developed, in which the requirement for a fixed ratio of turbocharged gasoline and diesel to hybrid electric powertrains was relaxed. This final scenario relies heavily on gasoline-fueled HEVs, which offer the greatest fuel consumption benefit relative to the other powertrains. In the Scenario IV analysis, slightly more than half of new vehicles sold were assumed to be HEVs. The remaining new vehicles were divided evenly between naturally-aspirated gasoline, turbo-charged gasoline, and diesel vehicles. Vehicle weight was reduced by 20 percent, mostly achieved with the use of lightweight materials, while the new vehicle fleet's size distribution remained unchanged. Vehicle acceleration performance improved slightly from today. The average new car accelerates from zero to 100 km per hour in 8.1 seconds, and the light truck does the same in 8.8 seconds. So when a high percentage of hybrids, roughly 55 percent, are relied on to achieve most of the fuel consumption reduction, less weight and size reduction was required to achieve modest improvements in vehicle performance.

All four scenarios reveal that achieving a factor-of-two reduction in fuel consumption by 2035 is possible, but requires aggressive action beginning today. The following sections compare the four scenarios on the basis of material cycle energy and GHG emissions impact, and their cost-effectiveness.

Material Cycle Impact Assessment

The material cycle refers to the energy and environmental impact of producing the materials embodied in the vehicles. It includes the material extraction and processing steps, but does not include transportation of the materials, or manufacturing and assembly of the vehicle. All the scenarios involve use of alternative lightweight materials and HEVs with lithium-ion batteries, each of which require greater amounts of energy and GHG emissions to produce, relative to today's conventional naturally-aspirated gasoline vehicle.

The material production impact of these changes was calculated by keeping track of the material composition of future vehicles, and the energy intensity of these materials. Energy intensity data was obtained from Argonne National Laboratory's Greenhouse gases, Regulated Emissions, and Energy use in Transportation (GREET 2.7) model. The two metrics compared across the scenarios are the energy consumed and metric tons of CO_2 emitted during the material cycle. The results obtained are reported in Table 4.5.

All four scenarios that halve the fuel consumption of future new vehicles resulted in higher energy use and CO_2 emissions during the material production phase, mainly due to use of more energy intense lightweight materials. The production energy requirement of primary aluminum, for example, is about five

Table 4.5 Material cycle impact of the average new car and of the new vehicle fleet in 2035

Scenario	Material cycle impact per gasoline car		Total material cycle impact of the new vehicle fleet	
	Energy (GJ/veh)	CO_2 emissions (ton/veh)	Energy (EJ)	CO_2 emissions (mil tons)
2006	88.2	6.80	1.78	137
I	92.1	6.90	2.35	176
II	97.1	7.34	2.51	189
III	91.8	6.88	2.37	177
IV	97.7	7.38	2.54	190

times that of the primary steel it replaces in the future lightweight vehicle. Even so, a model year 2035 car that consumes half the fuel of today's car will end up using 43 percent less energy over its lifetime, since the material cycle is responsible for only 10 percent of the vehicle's total lifecycle energy use and GHG emissions today. This analysis included the vehicle's material cycle, manufacturing and assembly, use phase, and end-of-life treatment in its life-cycle, but excluded energy demand to produce the fuel.

The calculated material cycle impact was not very different across the scenarios. The total energy consumed in producing materials embodied in new vehicles is about 2.3–2.5 exajoules (EJ) and GHG emissions in the form of CO_2 ranged from 175 to 190 million metric tons. Scenarios II and IV included the heaviest vehicles, and, therefore, they showed higher material cycle impacts, since they embody more materials than in the other scenarios.

Cost Assessment

Implementing improvements and new technologies to reduce fuel consumption will increase the cost of producing vehicles, and in turn, the retail price paid by consumers. The next stage in the analysis evaluated the cost of halving the fuel consumption of new vehicles in 2035, and compare this against the resulting savings in fuel use and GHG emissions.

The MIT researchers developed estimates of the additional production cost of improvements in future vehicles from a literature survey of future technology assessments (DOT, 2006b; EEA, 2002; NRC, 2002; NESCCAF, 2004; TNO, 2006, IEEP, LAT, 2006; Weiss et al., 2000). Production costs were assumed to account for all of the costs associated with producing a vehicle at the manufacturing plant gate. This included vehicle manufacturing, and corporate and production overhead. It excluded distribution costs and manufacturer and dealer profit margins (Vyas et al., 2000). The average cost of a naturally-aspirated gasoline vehicle today was assumed to be $14,000 for cars and $14,500 for trucks. All costs were given in 2007 U.S. dollars. Base costs of naturally-aspirated gasoline vehicles were taken to be the U.S. base retail price

Table 4.6 Increase in cost relative to a current naturally aspirated gasoline vehicle

Vehicle technology	Assumptions	Cost increase	
		Cars, US$2007	Light trucks, US$ 2007
2035 N.A. Gasoline	Engine and transmission improvements; engine downsizing and 20% weight reduction; reduced drag and rolling friction	$1,400	$1,600

of a Toyota Camry CE mid-size sedan and Ford F150 pickup truck, reduced by a factor of 1.4.

Improvements in engine, transmission, rolling friction and drag are expected to occur over the next three decades. If there is a strong emphasis on reducing fuel consumption, these improvements will occur alongside weight reduction and engine downsizing. As shown in Table 4.6, the cost of a 2035 naturally-aspirated gasoline car is estimated to increase by $1,400, and trucks by $1,600, relative to current vehicles, given the emphasis on reducing fuel consumption in the future.

Alternative powertrains and further weight reduction can lower fuel consumption further at additional cost. As shown in Table 4.7, it is estimated that turbocharging a 2035 gasoline car would cost an extra $500, bringing the total cost of a turbocharged 2035 car to $14,000 + $1,400 + $500 = $15,900. Table 4.8 shows the cost estimates assumed for each type of weight reduction. The percentage reductions for each of the weight reduction methods shown in this table were combined multiplicatively.

Weight reduction by material substitution is estimated to cost $3 per kilogram up to a 14 percent reduction in vehicle weight, and is accompanied by an additional 7 percent weight reduction from vehicle redesign and component downsizing that is cost neutral. Multiplicatively combining these reductions yields a 20 percent reduction in vehicle weight, which is equivalent to the

Table 4.7 Additional cost relative to a 2035 naturally-aspirated gasoline vehicle

Vehicle technology	Assumptions	Additional cost relative to 2035 N.A. gasoline vehicle	
		Cars, US$ 2007	Light trucks, US$ 2007
Alternative Powertrains			
2035 Turbocharged gasoline	Turbocharged spark-ignition gasoline engine	$500	$600
2035 Diesel	High-speed, turbocharged diesel; meets future emission standards	$1,200	$1,500
2035 Hybrid gasoline	Full hybrid; cost includes electric motor, Li-ion battery	$1,800	$2,300

Table 4.8 Estimated costs of vehicle weight reduction relative to a 2035 naturally-aspirated gasoline vehicle

Type of weight reduction	% vehicle weight reduction [%]	Additional cost relative to a 2035 N.A. gasoline vehicle [US$ 2007/kg]
First tier material substitution	14	$3
Component downsizing, vehicle redesign	7	$0
Subtotal	20	$2
Second tier material substitution	7	$5
Component downsizing, vehicle redesign	3	$0
Subtotal	10	$3.5
Vehicle size reduction	10	$0
Total	35	$2

reduction assumed for full emphasis on reducing fuel consumption, in other words 100 percent ERFC, at a cost of roughly $3.5 per kilogram. A second tier of more costly material substitution can yield an additional 7 percent reduction in vehicle weight at an estimated cost of $5 per kilogram, enabling an extra 3 percent reduction from further cost-neutral redesign and component downsizing. Finally, an additional 10 percent reduction is available by reducing the average size of vehicle the vehicle fleet. While size reduction is assumed to be cost neutral with respect to production costs, shifting to smaller vehicles implies some qualitative costs to the consumer from forgone interior volume. Multiplicatively combining these reductions yields a 35 percent total reduction in vehicle weight at an overall cost of roughly $2 per kilogram.

Given these cost estimates, the benefits of the different technology options can be compared by calculating the gross cost of reducing one metric ton of GHG emissions, expressed in dollars per ton of CO_2 equivalent ($/ton CO_2e), as shown below. The gross cost does not account for the value of fuel savings generated from lower fuel consumption when calculating the cost reducing of GHG emissions.

$$Cost\ of\ reducing\ one\ ton\ of\ GHG\ emissions$$
$$= \frac{Cost\ of\ reducing\ fuel\ consumption\ (FC)}{GHG\ emissions\ savings}$$

The cost of reducing fuel consumption is the sum of the cost of incremental improvements to conventional vehicle technology that reduce fuel consumption, plus any extra cost for upgrading to an alternative powertrain and adding lower weight components. The cost of incremental improvements in conventional vehicle technology that lower fuel consumption is estimated by multiplying the extra cost of the 2035 naturally-aspirated gasoline vehicle relative to today by the emphasis on reducing fuel consumption, measured as the percent

Table 4.9 The cost of reducing one ton of GHG emissions in 2035 cars and light trucks

Vehicle technology	Gross cost of GHG reduction, in US$ 2007/ton CO_2e		Undiscounted payback period, in years	
	Cars	Light trucks	Cars	Light trucks
N.A. Gasoline	55	50	4	4
Turbocharged Gasoline	60	55	4	4
Hybrid Gasoline	70	70	5	5
Diesel	80	70	6	5

ERFC. It was assumed that the efficiency gains provided by changing to an alternative powertrain, or by additional weight reduction, are fully realized in lowering fuel consumption. The remaining portion of the 2035 naturally-aspirated gasoline vehicle cost is attributed to other benefits, such as increasing size, weight or improving performance.

It was assumed that all of the efficiency improvements in conventional vehicle technology are directed towards reducing fuel consumption and that vehicle weight is reduced by 20 percent between today and 2035. GHG emissions savings are calculated relative to what they would be if the fuel consumption of a 2035 vehicle remains unchanged from 2006, assuming a lifetime vehicle travel of 240,000 km over 15 years. Data on the average of lifetime car and light truck travel were obtained from the U.S. Department of Transportation vehicle survivability and mileage travel schedule (DOT, 2006a). The results of applying this approach are shown in Table 4.9.

The estimated gross cost of reducing GHG emissions ranges from $50 to $80 per ton CO_2e, yielding a variation in cost of roughly 50 percent across an average of $65 per ton CO_2e. An improved 2035 naturally-aspirated gasoline vehicle realizes the most cost-effective reductions in GHG emissions and fuel use when all future efficiency improvements are realized in reduced fuel consumption. In cars, diesel engines are less cost effective than turbocharged or hybrid electric powertrains, but in trucks, diesel engines are about as cost effective as hybrid electric drivetrains. Assuming a constant fuel cost of $1.85 per gallon, the value of the undiscounted fuel savings recoups the initial gross cost of each of the different vehicle technologies within 4–6 years. The fuel price of $1.85 per gallon was taken as the average of the Energy Information Administration (EIA) "Annual Energy Outlook" long-term forecast for motor gasoline, excluding $0.40 per gallon in federal, state, and local taxes (EIA, 2007b).

The results in Table 4.9 have embedded a 20 percent reduction in vehicle weight by 2035. When separated out from the alternative powertrain and other vehicle improvements, weight reduction on its own has an estimated gross cost between $75 and $80 per ton CO_2e for cars, and between $65 and $70 for trucks. Thus, while reducing vehicle weight realizes extra savings in fuel use and GHG emissions, these benefits come at a higher marginal cost that raises the cost of reducing a ton of CO_2 overall, although these costs are still recouped within 5–6 years by the value of the fuel savings generated from reducing vehicle weight.

Table 4.10 Societal costs, benefits, and cost-effectiveness of halving fuel consumption in 2035 model year vehicles across the four scenarios (all values in 2007 U.S. dollars)

Scenario	Extra cost to halve Fuel consumption of 2035 model year vehicles (billion $US)	As % of baseline cost	Undiscounted fuel savings pay-back period, in years	Gross cost of GHG reduction, $US/ton CO_2e
I	$54	16	4	$65
II	$56	17	5	$70
III	$63	19	5	$76
IV	$58	17	5	$72

Next, the results from Table 4.9 were extrapolated across all new vehicles in 2035 to develop an estimate of the total societal costs of halving fuel consumption of the 2035 model year. Table 4.10 shows the aggregate extra cost of all new 2035 model year vehicles in each of the three bounding scenarios that halve new vehicle fuel consumption by 2035. Over 15-year lifetimes, vehicles in the 2035 model year will save 290 billion liters of fuel and offset a total of 850 million tons of GHG emissions. This is roughly equivalent to half of the total of motor gasoline fuel used in the U.S. in 2006 (EIA, 2007a).

The extra cost of halving fuel consumption shown in Table 4.10 is the combined cost of all efficiency improvements necessary to halve fuel consumption in new vehicles in 2035. Depending on the scenario, the extra cost ranges from $54 to $63 billion. This is equivalent to an additional 16–19 percent of the estimated baseline production cost of the 2035 model year when average fuel consumption remains unchanged from 2006. Assuming a 15-year lifecycle, a fuel cost of $1.85 per gallon, and a discount rate of 3 percent, the value of the fuel savings provided by vehicles in the 2035 model year is estimated at $120 billion, which would yield a total net societal gain of some $60–$70 billion, after subtracting the extra costs of halving fuel consumption. The 3 percent discount rate is the same as the "social rate of time preference" used by the U.S. Office of Management and Budget (OMB, 2003). The undiscounted payback period to recoup the initial extra cost of halving fuel consumption is roughly 4–5 years.

These estimates do not take into account the rebound effect of increased vehicle travel as it becomes cheaper to drive a vehicle with lower fuel consumption. Most studies have placed the long-term rebound effect between 10 and 25 percent (Greening et al., 2000). Small and Van Dender (2005) however, recently found that between 1997 and 2001, the long-term rebound effect was half of its value over the entire 1966–2001 period, and is likely to diminish below 10 percent as rising income reduces the relevance of fuel costs in travel decisions.

Without accounting for fuel savings, the cost of reducing a ton of GHG emissions ranges from $65 to $76 across the three scenarios. For comparison, the Intergovernmental Panel on Climate Change (IPCC) estimates that GHG reductions costing between $20 and $80 per ton of CO_2e before 2030, and between $30 and $150 by 2050, will be required in order to stabilize atmospheric GHG emissions at 550 parts per million CO_2e by 2100 (IPCC, 2007).

Conclusions

The MIT analysis has examined the necessary changes required to double the fuel economy, or halve the fuel consumption of new vehicles within the next three decades. The results reveal the following key conclusions:

- Technologies are available to do the job. With the set of light-duty vehicle options deployable in the nearer term, it is possible on average to halve the fuel consumption of new vehicles by 2035. This requires improvements in the engine and transmission; aerodynamic drag, rolling resistance and weight and size reduction; and deployment of more efficient alternative powertrains.
- Significant changes are required, and there are trade-offs. The MIT study reveals important trade-offs between the performance, cost, and fuel consumption reduction benefits. For example, Scenario I is the most cost effective, but leaves performance at today's levels. Conversely, Scenario III offers the largest performance improvement of all scenarios presented, but is more expensive and requires aggressive weight reduction and use of alternative powertrains. Costs are higher for scenarios that direct future efficiency improvements towards increasing vehicle horsepower and acceleration performance, rather than towards reducing fuel consumption.
- The production cost of future vehicles will rise. Halving the fuel consumption by year 2035 will increase the production cost of future vehicles with roughly the same size, weight, and performance as today. Excluding distribution costs and dealer and manufacturer profits, the total extra cost of the 2035 model year vehicles is estimated at $54–$63 billion, or about 20 percent more than the baseline cost. This corresponds to a cost of $65–$76 per ton of CO_2e emissions avoided, when accounting for emissions savings over the lifetime of vehicles in the 2035 model year.

It turns out that the new U.S. national fuel economy standards passed near the end of 2007 seek fuel economy improvements that are similar to the factor-of-two goal that has been evaluated. Under the new legislation, the industry is required to achieve a CAFE rating of 35 mpg by 2020. Like the factor-of-two target, this requires the average fuel economy to increase at a compounded rate of 3 percent per year; thus similar conclusions can be drawn regarding the situation now facing the automotive industry.

The new CAFE rule adopted in December 2007 is worthy in its objective of reducing the impact of future light-duty vehicles related to energy security and global warming. While it is technically possible to meet the new standards, the nature and magnitude of the changes required run counter to the recent trends towards larger, heavier, more powerful vehicles. The scenarios evaluated in this chapter depict a transportation future where automakers face costs up to 20 percent higher to produce potentially smaller vehicles with performance similar to today's vehicles. The future will challenge the auto industry to make the capital investments necessary to realize more efficient technologies at a substantial scale. It will require the government to address the market failures that promote size, weight, and acceleration or horsepower performance at the expense of higher vehicle fuel consumption.

These are striking changes from the status quo. Meeting the new CAFE standards or halving fuel consumption in 2035 vehicles will each require a fundamental shift in the mindset and motivation of consumers, industry, and governmental stakeholders. Automakers may be hesitant to make such large-scale changes in the product mix unless consumers indicate they are willing to forego their continuing pursuit of ever higher performance, larger vehicle size, and other amenities. A set of policies complementary to the CAFE program that stimulate market demand for lower fuel consumption might put us in a better position to achieve this worthy and ambitious goal.

References

Energy and Environmental Analysis Inc. (EEA), Analysis and Forecast of the Performance and Cost of Conventional and Electric-Hybrid Vehicles, (2002). Accessed 19 February 2007, from http://www.energy.ca.gov/fuels/petroleum_dependence/documents/2002-04-09_HYBRID. PDF

Energy Information Administration (EIA), Short-Term Energy Outlook – July 2007, (2007a). Accessed July 2007 from http://www.eia.doe.gov/emeu/steo/pub/contents.html

Energy Information Administration (EIA), Annual Energy Outlook 2007 with Projections to 2030, Report #DOE/EIA-0383, (2007b). Accessed July 2007 from http://www.eia.doe.gov/oiaf/aeo/index.html and http://www.eia.doe.gov/oiaf/aeo/supplement/supref.html#trans, Tables 47, 58

Greening, L. et al., Energy efficiency and consumption – the rebound effect – a survey, Energy Policy, 28 (2000) pp. 389–401

Heavenrich, R., Light-Duty Automotive Technology and Fuel Economy Trends: 1975 through 2006, (2006) U.S. Environmental Protection Agency, EPA420-S-06-003, July 2006

IPCC (2007), Climate Change 2007: Mitigation of Climate Change. Working Group III contribution to the Intergovernmental Panel on Climate Change Fourth Assessment Report: Summary for Policy Makers, para. 23, p. 29. Accessed July 2007 from http://www.mnp.nl/ipcc/pages_media/ar4.html

Kasseris, E. and J. B. Heywood, Comparative Analysis of Automotive Powertrain Choices for the Next 25 Years, (2007) SAE Paper No. 2007-01-1605, SAE World Congress & Exhibition, Detroit, MI, USA, April 2007

Kromer, M. A. and J. B. Heywood, Electric Powertrains: Opportunities and Challenges in the U.S. Light-Duty Vehicle Fleet, Laboratory for Energy and the Environment report #LFEE 2007-03 RP, (2007) Massachusetts Institute of Technology, June 2007, Accessed August, 2007 from http://lfee.mit.edu/public/LFEE_2007-03_RP.pdf

National Research Council, Board on Energy and Environmental Systems, Effectiveness and Impact of Corporate Average Fuel Economy (CAFE) Standards, (2002) National Academy Press, Washington, D.C.

Northeastern States Center for a Clean Air Future (NESCCAF), Reducing Greenhouse Gas Emissions from Light-Duty Motor Vehicles, 2004. (2004) Accessed July 2007 from http://www.nescaum.org/documents/rpt040923ghglightduty.pdf/

Office of Management and Budget (OMB), Circular No. A-4: Regulatory Analysis, (2003). Accessed August 2007 from http://www.whitehouse.gov/omb/circulars/a004/a-4.pdf

Small, K. A. and K. V. Dender. The effect of improved fuel economy on vehicle miles traveled: Estimating the rebound effect using U.S. state data 1966–2001 (2005). *Working paper no. 05-06-03* Retrieved 31 July 2007, from http://repositories.cdlib.org/cgi/viewcontent.cgi?article = 1026&context = ucei

TNO Science and Industry, Institute for European Environmental Policy (IEEP), & Laboratory of Applied Thermodynamics (LAT), Review and analysis of the reduction potential and costs of technological and other measures to reduce CO2-emissions from passenger cars, 2006. (2006) Retrieved 2 March 2007, from http://www.lowcvp.org.uk/assets/reports/TNO%20IEEP%20LAT%20et%20al%20report_co2_reduction.pdf

U.S. Department of Transportation (DOT), Vehicle Survivability and Travel Mileage Schedules, National Highway Traffic Safety Administration, Technical Report DOT HS 809 952, (2006a). Accessed July 2007 from http://www-nrd.nhtsa.dot.gov/pdf/nrd-30/NCSA/Rpts/2006/809952.pdf January 2006

U.S. Department of Transportation (DOT), Final Regulatory Impact Analysis: Corporate Average Fuel Economy and CAFE Reform for MY 2008-2011 Light Trucks, 2006. (2006b) Accessed September 29, 2006, from http://www.nhtsa.dot.gov/staticfiles/DOT/NHTSA/Rulemaking/Rules/Associated%20Files/2006_FRIAPublic.pdf

Vyas, A., D. Santini, R. Cuenca, Comparison of indirect cost multipliers for vehicle manufacturing, Argonne National Laboratory, (2000). Accessed July 2007 from http://www.transportation.anl.gov/pdfs/TA/57.pdf April 2000

Weiss, M. A. et al., On the Road in 2020: A life-cycle analysis of new automobile technologies, (2000), Cambridge, United States: Laboratory for Energy and the Environment (LFEE). Accessed July 2007 from http://web.mit.edu/energylab/www/pubs/el00-003.pdf

Chapter 5
Lead Time, Customers, and Technology: Technology Opportunities and Limits on the Rate of Deployment

John German

The automotive industry is in a period of unprecedented technology development that will move the world's transportation system a long way towards "sustainable mobility." Gasoline engine technology is maturing rapidly and manufacturers are working hard on diesel engines suitable for use in light duty vehicles, including conventional passenger cars and light duty trucks (LDTs), that can meet United States (U.S.) emission standards. Honda is already producing third generation hybrid electric vehicles (HEVs) and most other manufacturers have or will be introducing HEVs. Honda continues to market a dedicated compressed natural gas vehicle, the Civic GX, and a number of manufacturers produce flexible-fuel vehicles that run on gasoline or mixtures containing 85 percent ethanol (E85). Fuel cells are being heavily researched and developed. All of these vehicles, displayed in Fig. 5.1, achieve carbon dioxide (CO_2) reductions compared to conventional gasoline vehicles.

Development of automotive technologies is accelerating in response to growing concerns about energy security and global warming. Demand for transportation energy is so immense that no single technology can possibly be the solution. Rapid development and implementation of as many feasible technologies as possible is needed. This chapter discusses technology development for cars and LDTs that meet the needs of customers and the global need to address climate change and energy sustainability.

Conventional Technologies

Gasoline engines will continue to dominate in the light duty vehicle marketplace for some time. Future gasoline engines will be much more efficient than current engines. Improvements that are currently being implemented include variable valve timing, direct fuel injection, variable cylinder shutoff, and smaller displacement turbocharged engines. Honda pioneered variable valve timing and

J. German
American Honda Motor Company, 3947 Research Park Drive, Ann Arbor MI 48108, USA

D. Sperling, J.S. Cannon (eds.), *Reducing Climate Impacts in the Transportation Sector*, DOI: 10.1007/978-1-4020-6979-6_5, © Springer Science+Business Media B.V. 2009

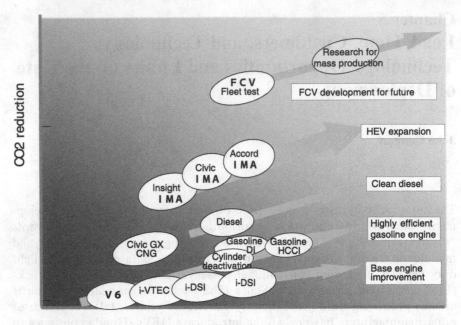

Fig. 5.1 Honda's powertrain progress for CO2 reduction

lift (VTEC) and now installs it on all of its vehicles. This system allows switching from a low lift/duration camshaft, which optimizes emissions and fuel economy during normal driving, to a high lift/duration camshaft which increases performance when needed. For the 2008 model year, Honda introduced an improved variable cylinder management system on most Honda Accord V6 and Honda Odyssey models that switches from 6-cylinder to 4-clyinder or 3-cylinder operation, depending on vehicle load.

Computer controls are also enabling a variety of improved transmission designs. The dual-clutch automated manual transmission has very smooth shifts without any torque interruption, is almost as efficient as a manual transmission, and is potentially less expensive. However, the lack of a torque converter makes it more difficult to launch from a stop and it requires huge investments to completely retool transmission production. The continuously variable transmission (CVT) is extremely smooth and allows the engine to run at maximum efficiency during urban driving. It can also deliver steady-state engine speeds to facilitate homogeneous charge compression ignition (HCCI) engine operation. However, it is torque limited which prevents use on larger engines, has high belt friction which limits efficiency on the highway, and also requires huge investments. Much of the efficiency gains can be achieved on conventional automatic transmissions at a fraction of the investment cost by improving shift points and lock-up strategies, adding additional gears, and moving to lepelletier 6- to 8-speed automatics with fewer clutches and planetary

Table 5.1 Incremental fuel economy technology

Engine technology
- High specific output (including 4 valve/cylinder)
- Variable valve timing/lift
- Cylinder deactivation
- Direct injection
- Precise air/fuel metering
- Lower engine friction
- Turbocharging

Transmission efficiency
- 5/6/7/8 speed
- CVT
- Duel-clutch automated MT

Reduced losses
- Lightweight materials
- Low drag coefficient
- Low resistance tires
- Lower accessory losses

gear sets. It is not yet clear which strategy is the most cost-effective for different vehicles and all will likely co-exist for at least a decade or two.

Table 5.1 provides a partial list of technologies being used to improve the efficiency of conventional gasoline engines. These technologies are continuously being incorporated into vehicles, but usually not to improve fuel economy. These technologies are being used to improve other attributes valued more highly than fuel economy by most consumers, such as performance, safety, utility, and luxury. The real challenge to improving fuel economy is not technology, but getting customers to accept use of the technology to improve fuel economy instead of more highly valued attributes.

Diesels

Light duty diesel engines enjoyed a brief sales surge in the United States in response to the two oil crises in the 1970s. Unfortunately, these diesels were slow, loud, rough, difficult to start, distinctively malodorous, and, in some cases, unreliable. As fuel prices dropped and nitrogen oxides emission standards became more stringent, diesel car models all but disappeared in the United States.

The situation was very different in Europe, however, where high fuel taxes in general and much lower taxes on diesel fuel compared to gasoline in most countries spurred development of vastly improved diesel engines, enabled by computer controls. Europe also established less stringent emission standards for diesels, allowing them to emit nitrogen oxide pollution at levels an order of magnitude higher than in the United States. As a result, Europe currently sells

millions of vastly improved light-duty diesels each year with excellent performance, although at a substantial cost increment.

Meeting the U.S. Tier 2 bin 5 nitrogen oxides standard set by the Environmental Protection Agency is a major challenge for light-duty diesel engines.
Engine-out emissions are lower in diesels than gasoline engines, but nitrogen
oxides cannot be reduced in conventional 3-way catalysts during the lean
operation inherent to the diesel engine. Thus, manufacturers have been developing new techniques to trap emissions or introduce a nitrogen oxide reactant
while the diesel is running lean.

Most manufacturers are investing in urea injection. Urea is stored onboard
the vehicle in a separate tank and is injected before the catalyst to provide a
reactive agent that reduces nitrogen oxides while the diesel is running lean. The
process is referred to as selective catalytic reduction (SCR). This process is
effective and a number of manufacturers have announced plans to produce
diesels meeting the Tier 2 bin 5 emission standards for 2008 and 2009 using urea
injection.

Honda is concerned about the additional complexity of the urea system
and the inconvenience to customers of refilling the urea tank. Also, if solutions
other than urea injection become widespread in the future, the owners of urea-
equipped vehicles may some day have difficulty finding urea to operate their
vehicles. Thus, Honda has developed a new approach that traps nitrogen oxides
and also creates a SCR process without the need for urea injection. This process
is illustrated in Fig. 5.2:

1. During the initial lean burn operation event, the adsorbent in the catalyst
   captures nitrogen oxides from the exhaust gas.
2. When needed, the engine management system adjusts the engine air/fuel
   ratio to rich-burn operation. The nitrogen oxides in the adsorption layer

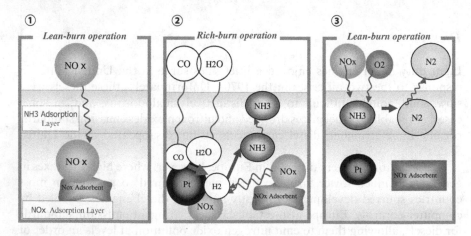

Fig. 5.2 Honda catalyst for Tier 2 Bin 5 diesel

react with hydrogen obtained from the exhaust gas to produce ammonia and harmless nitrogen. Another adsorbent material temporarily adsorbs the ammonia.

3. When the engine returns to lean-burn operation, the ammonia adsorbed during the rich event reacts with nitrogen oxides in the exhaust gas and reduces them to nitrogen through the SCR process. Nitrogen oxides are also adsorbed in the catalyst, as in step 1.

Hybrid Electric Vehicles

HEVs have a number of attractive features, in addition to much lower fuel costs. They have the best possible "idle" quality, as no other vehicle can match shutting the engine off in HEVs for smoothness and quiet. Some customers intensely dislike going to service stations and greatly value the superior driving range of HEVs. There is also pride in demonstrating one's commitment to benefiting society. The electric motor provides almost all of its power at zero revolutions per minute (rpm), resulting in a large torque boost at low speeds and a very strong launch assist. In fact, as illustrated in Fig. 5.3, the torque curve of a Honda Civic Hybrid is very similar to that of a diesel engine.

Honda introduced the first HEV in the United States, the Insight, in 1999 and followed up with hybrid electric versions of the popular Civic in 2002, which was redesigned in 2005, and the Accord in 2004. Honda is currently developing an all-new, dedicated, more affordable HEV. This car, which will be

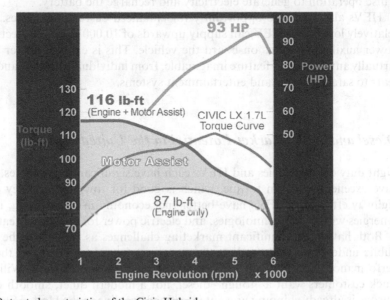

Fig. 5.3 Output characteristics of the Civic Hybrid

available only with a hybrid electric drivetrain, will be launched in North America in 2009, and will cost significantly less than the Civic Hybrid. The North American sales volume target is 100,000 units per year.

The primary challenge for HEVs is to reduce the cost of the system. Cost reductions should gradually occur with increased sales, more suppliers, and further development. Sales will increase as costs come down, leading to more cost reductions.

HEVs also have a number of synergies with other technologies that can increase the benefits of the system. Electric pumps and compressors are more efficient than mechanical ones and eliminate accessory belts, resulting in lower maintenance and less engine space needed in vehicle design. Part-time four wheel drive systems using an extra electric motor are less expensive and more efficient than conventional all-wheel drive systems. The assist from the electric motor can extend the operation window for cylinder deactivation, as on the Accord Hybrid, and make up for the loss of power with Atkinson cycle operation, as on the Prius HEV. Over the long term, the motor assist can provide quasi-steady-state load conditions to help enable HCCI operation, especially when combined with a continuous variable transmission to maintain constant engine speed. Finally, an electric motor can be placed in-line with the turbocharger compressor, often referred to as an e-turbo. These systems are currently available, but, with conventional 12 Volt systems, the motor is only capable of spinning the turbocharger up a little faster to reduce turbo lag. With high voltage electric power from the hybrid electric system, however, the e-turbo could provide full supercharger boost in addition to turbocharger boost. There is also the potential to use exhaust energy to drive the e-turbo motor during cruise operation to generate electricity and recharge the battery.

HEVs also offer the opportunity for enhanced customer features. Even a relatively low power HEV can supply upwards of 10,000 Watts of electricity to power auxiliary systems on-board the vehicle. This is enough power to offer virtually any consumer feature imaginable, from individually heated and cooled seats to safety systems and entertainment systems.

Diesel and HEV Market Potential in the United States

Light duty diesel vehicles and HEVs each have significant advantages. Diesels have excellent low-rpm torque, which is good for towing, and they excel in highway efficiency. HEVs have better fuel economy in city driving, multiple synergies with other technologies, and electric power for consumer features.

Both have some significant marketing challenges as well. Will the general public understand that modern diesel engines have largely solved the noise, performance, vibration, smell, and starting problems of the past? Will pickup truck customers want a "tough" diesel, not a modern quiet, smooth one? As Europe is already shipping unwanted, refined gasoline to the United States, can

U.S. refineries adjust their diesel and gasoline output if the United States also shifts to diesel? Will mainstream customers believe that HEV batteries will not have to be replaced and that the hybrid electric drivetrains can be easily serviced?

While these are legitimate concerns, the primary challenge for both diesels and HEVs is simply cost. Diesels are currently cheaper than HEVs, but they are not cheap. In a 2004 report, Greene, Duleep, and McManus estimated that the incremental cost for a diesel meeting Euro IV emission standards over a gasoline engine is $1,700 for a 4-cylinder and between $2,300 and $2,500 for a V-6 (Greene et al., 2004). These incremental costs will increase substantially as equipment is added to meet EPA Tier 2 emission standards, while HEV costs will come down with further market penetration and development. In the short term, both will appeal to significant market segments and will increase market share. Diesels and HEVs will likely appeal to different markets, with diesels favored for larger vehicles and rural areas that do mostly highway driving and hybrids favored for smaller vehicles and urban areas.

Both diesels and HEVs are currently too expensive to be accepted by main-stream customers. The key factor in their long-term market share will be cost reduction. Diesels face an additional challenge from next-generation gasoline engines, as discussed below. The benefits from hybridization will be much less affected by gasoline engine improvements, so HEVs will gradually increase market share as cost comes down and synergies increase.

Next-Generation Gasoline Engines

Improvements to the gasoline engine are currently in development. Two of the most significant technologies are camless valves and HCCI. Camless valves would replace the camshaft with individual valve actuators that can be controlled by the computer. Figure. 5.4 illustrates an electrical-mechanical system, although other designs are possible. The primary advantage of camless valves is that they allow optimization of the combustion cycle to varying speed and load conditions.

HCCI has a much higher heat release rate than conventional spark ignition, as shown in Fig. 5.4, which leads to much lower heat losses and higher efficiency. However, because ignition is controlled only by the compression of the air/fuel mixture, HCCI is far more difficult to control than conventional spark ignition or diesel ignition, which is controlled by fuel injection timing. Camless valves would likely help in creating the conditions necessary for HCCI operation, although some limited use of HCCI in gasoline engines is likely without camless valves.

Figure 5.5 illustrates one possible advanced gasoline scenario. This should not be considered a prediction, as it is impossible to predict how technologies will develop and what combination of technologies will eventually prove to be

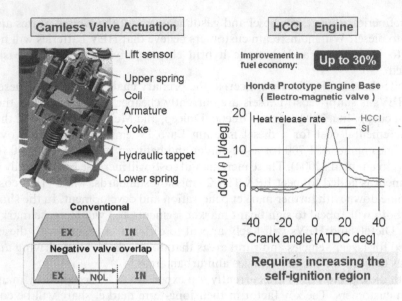

Fig. 5.4 Next-generation gasoline engines

most cost effective. It should also not be considered a short-term scenario, as many of the technologies are still in the early stages of development and the integration of the different technologies will require lengthy, iterative implementation. Rather, it is presented to illustrate what the diesel and other advanced technologies are up against in the 25–30 year time frame. The combustion strategy in Fig. 5.5 assumes the development and use of camless valves, direct fuel injection, e-turbo, a relatively small hybrid electric system, and a severely downsized engine.

At low speeds and at low loads when the engine is least efficient, the engine is shut off and the vehicle operates on the electric motor. When conditions are suitable, HCCI operation is used. The engine is boosted to extend the

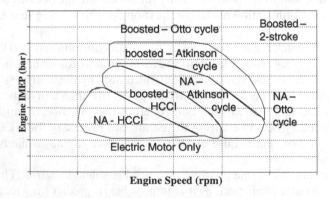

Fig. 5.5 Potential engine operation modes

HCCI operating window, with boost available from whatever combination of supercharger, turbocharger, and electric motor proves to be best. Once HCCI operating conditions are exceeded, the engine switches to Atkinson cycle operation. This has higher efficiency than the conventional Otto cycle, but not as high as HCCI. Again, the Atkinson cycle operating window is extended with boost from the e-turbo and electric motor. For high power, the engine switches to Otto cycle or runs as a two-stroke engine for maximum power. The combination of two-stroke operation with supercharger, turbocharger, and electric motor boost would allow the engine to be roughly a third the size of current engines, reducing engine friction and further boosting efficiency. Finally, the battery pack can be recharged both from regenerative braking and from capturing exhaust energy with the e-turbo.

The advanced engine alone, without the hybrid electric system, will likely be almost as efficient as an advanced diesel engine, at substantially lower cost. Cheah et al., have estimated that the cost of a diesel engine in 2035 would be $700–$900 more than the cost of an advanced gasoline turbocharged engine (Cheah et al., 2007). This will likely limit the market share of the diesel engine in the long run. Any advanced technology vehicle needs to be compared to such a theoretical engine, not current gasoline engines. The future gasoline engine will greatly raise the bar that must be cleared by alternative powertrains.

Alternatives to Petroleum

Alternative fuels, including advanced plug-in hybrid electric vehicles (PHEVs) that rely on electricity from the grid for part of their traction power, provide other opportunities to reduce carbon dioxide emissions and petroleum consumption from cars and LDTs.

Civic GX Natural Gas Vehicle

The natural gas powered Honda Civic GX, now in its third generation, offers multiple benefits, including about a 20 percent reduction in CO_2 emissions on a gallon equivalent basis, near-zero emissions, high reliability and durability, and home refueling cost savings and convenience. The current Civic GX has a real-world natural gas driving range of over 200 miles and is certified as a California advanced technology-partial zero emission vehicle (AT PZEV) and an EPA Tier 2 bin 2 inherent low emission vehicle (ILEV).

To provide more value to customers, Honda and Fuelmaker Corp. have introduced a home refueling appliance, called Phill, in California and New York. Phill is maintenance free, quiet, easy to use, and gives consumers the benefit of home refueling with natural gas that costs about one-third that of gasoline. Moreover, Honda believes that the application of compressed natural gas in the market can play a critical role in bridging the gap between gasoline and alternatives such as hydrogen, providing a mechanism for industry and government to gain a greater understanding of the challenges and opportunities associated with gaseous fuels and home refueling. In fact, Honda's first retail customer for the FCX fuel cell vehicle was selected from its existing base of Civic GX customers.

Plug-In Hybrid Electric Vehicles

The ability for PHEVs to use electricity for much of their energy has great potential to reduce petroleum use in the long run. The primary constraint is that PHEVs require a battery pack with at least five times the energy storage compared to the battery pack in a conventional power-split HEV. Current HEVs made by Toyota and Ford use a power-split design. Honda's HEVs use a simpler and less costly integrated electric motor system with a smaller battery pack. As the battery pack is already the single large cost of the hybrid system, this places a severe constraint on market acceptance of PHEVs.

The American Council for an Energy Efficient Economy (ACEEE) published a study in September 2006 assessing PHEV costs and benefits. (Kliesch and Langer, 2006) For its analysis, ACEEE assumed all vehicles were driven 12,000 miles per year and that gasoline prices were $3.00 per gallon. Baseline conventional vehicles were assumed to have an average fuel economy of 30 miles per gallon (mpg) and HEVs were assumed to attain a fuel economy of 50 mpg. The analysis assumed further that the PHEV batteries would last the life of the vehicle and that half the mileage driven by PHEVs was on electricity. No costs were assigned to the larger electric motor and power electronics needed for all-electric vehicle operation. Importantly, the study did not assess the impacts of the additional size and weight of the batteries, implicitly assuming no fuel economy or performance penalty for the additional weight of the batteries and no value for the loss of utility due to the additional size.

The first 3 columns of Table 5.2 present the results from ACEEE's analysis of a PHEV with a 40 mile all-electric-range (AER). The last column compares the PHEV to the HEV using the data from the 2nd and 3rd columns, which was not directly compared by ACEEE. The first case presents near-term incremental costs. This illustrates that at current battery costs, about $1,500 per kilowatt hour (kWh), the payback period—total incremental cost divided by annual fuel savings—for the PHEV is 27 years compared to a conventional vehicle and almost 69 years compared to the HEV.

The more interesting case is ACEEE's assessment of long-term incremental costs. ACEEE assumed that PHEV battery costs would drop by 80 percent to $300 per kWh. This is much lower than can be achieved by large-scale production of the newest lithium-ion battery designs, implying substantial improvement in basic battery technology. Even with this assumption, the PHEV

Table 5.2 Plug-in hybrid payback (ACEEE, Sep 2006)

	Hybrid	Plug-In, 40-mile range	Calculated Plug-In vs. hybrid
Near-term incremental costs			
Battery	$2,000	$17,500	$15,500
Other incremental costs	$1,500	$1,500	0
Annual fuel savings	$480	$705	$225
Payback (years)	7.3	27.0	68.9
Long-term incremental costs			
Battery	$600	$3,500	$2,900
Other incremental costs	$1,000	$1,000	0
Annual fuel savings	$480	$705	$225
Payback (years)	2.9	6.4	12.9

payback period is 6.4 years, similar to the current HEV payback of 7.3 years. This implies that, with an 80 percent reduction in battery cost and batteries that last the life of the vehicle, a PHEV niche market could develop similar to the current 2 percent market share for HEVs. However, note that with this level of battery cost reduction, the payback period for HEVs drops to just 2.9 years. This is within the two to three year payback period valued by most customers, implying that HEVs could be accepted by mainstream customers. Thus, the comparison should assume HEVs will be the base case, which yields a PHEV payback period of almost 13 years. This is still not a market case, even ignoring battery durability, size, weight, and recharging issues.

In reality, PHEV battery durability will be much shorter than batteries in conventional HEVs, due to the deep discharge cycles and higher loads at lower state-of-charge required for PHEV usage. Even if dramatic improvements in battery durability occur that allow PHEV batteries to last the life of the vehicle, application of these same durability improvements to conventional HEVs would enable a smaller, more robust, less expensive HEV battery, which still reduces PHEV cost effectiveness. Also, even advanced batteries will require an extra 4 cubic feet of space, reducing the vehicle utility and the customer's value of the vehicle. The battery pack will also add about 200 pounds, reducing the performance and fuel economy of the vehicle. Finally, PHEVs require a safe place to plug in. This will be difficult for second and third owners, which is likely to cause potential consumers to worry about resale value.

The current push for PHEVs as a near-term solution is no more feasible than the push was just a few years ago for near-term deployment of fuel cells. On the other hand, current petroleum consumption rates cannot be maintained or increased forever. Both PHEVs and fuel cells have great promise to replace petroleum use in the long run, but both need breakthroughs in order to support large-scale deployment.

Fuel Cell Vehicles

Honda has advanced its hydrogen fuel cell vehicle (FCV) technology though four generations of prototype vehicles and several generations of production models. It has focused on overcoming barriers to market acceptance by doing original technology research and development, and through early deployment of its technology with customers living in today's world. Honda's existing FCV technology delivers almost twice the tank-to-wheel fuel efficiency of its current gasoline fueled hybrid electric technology, with no harmful exhaust emissions.

The 2008 FCX model direction is a spacious sedan on a distinctive low floor design, made possible by an innovative and compact new Honda-developed V Flow fuel cell stack. The Honda V Flow stack improves the density of power to volume ratio by 50 percent and the power to weight ratio by 67 percent compared to the 2007 FCX. The new V Flow structure takes advantage of gravity to efficiently discharge water formed during electricity generation, improving performance in subzero temperatures and start-up at temperatures below –20°F. The fuel cell powertrain generates more power than the current FCX and has a 30 percent improvement in range, to about 280 miles. Hydrogen is stored in a 171 litre tank at 5,000 pounds per square inch.

Despite these potential benefits, creation, transport, and storage of hydrogen are significant barriers to hydrogen's viability as a mainstream alternative to gasoline. One approach to hydrogen refueling is the home energy station concept, shown schematically in Fig. 5.6. Natural gas is reformed to generate hydrogen for vehicle use. At the same time, some of the hydrogen is sent to a fuel cell inside the unit to generate electricity for the home. Heat, which is a byproduct of the electricity production process, is captured and used. The result is a total efficiency improvement. A second approach is renewable hydrogen production through Honda's own solar panel technology made of copper, indium, gallium and selenide, which reduces by half the

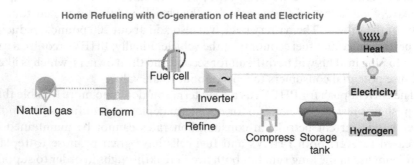

Home Refueling with Co-generation of Heat and Electricity

Fig. 5.6 Home energy station

energy and carbon dioxide emissions required for their manufacture. Mass production of the solar cells began in fall 2007.

Consumer Limits

One major challenge for manufacturers in introducing new technology is the low value placed by most consumers on fuel economy. Most customers understand that the real cost of driving is very low, and they value performance, utility, comfort, and safety more highly than fuel economy.

Real Cost of Driving

Figure 5.7 shows the real price of gasoline in the United States, corrected for inflation using the urban consumer price index. Even at $3.05 per gallon, gasoline prices are similar to what they were during the second oil crisis in the early 1980s and about 60 higher than they were before the first oil crisis.

However, that's not the whole story because the average fuel economy of the in-use vehicle fleet has improved substantially since 1980, as shown on Fig. 5.7. Figure 5.8 combines these effects and shows the average cost of driving a mile, adjusted for inflation. This is currently almost exactly the same as before the first oil crisis and significantly less than during either oil crisis.

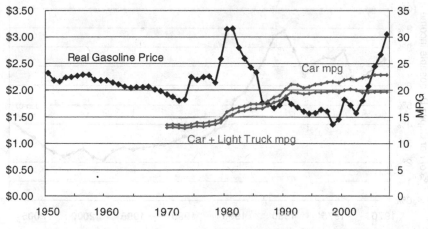

Motor Gasoline Retail Prices, U.S. City Average, adjusted using CPI-U
In-Use MPG from Transportation Energy Data Book: 2007

Fig. 5.7 Real gasoline price and in-use fleet MPG (2007 $ per gallon)

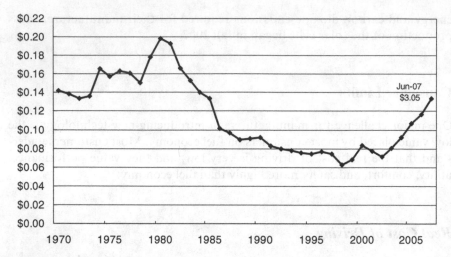

Fig. 5.8 Real gasoline cost for cars – cents per mile (2007 $ per gallon)

Even this is still not the whole story, because it doesn't account for substantial increases in our standard of living. As shown in Fig. 5.9, only about 4 percent of disposable income is now needed to drive 10,000 miles, compared to 9 percent during the second oil crisis and 6–7 percent before the first oil crisis. Gasoline would have to rise to about $5 per gallon before driving 10,000 miles would require the same percentage of disposable income as it did before the first oil crisis.

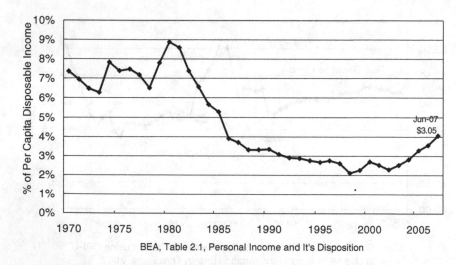

Fig. 5.9 Real fuel cost of driving a passenger car 10,000 Miles – % of per capita disposable income

Trade-Offs with Other Features Valued More Highly

The customer is an integral part of the equation. To help illustrate this, the graph on the left of Fig. 5.10 shows the changes in vehicle weight, performance, and proportion of automatic transmissions since 1981 in the passenger car fleet, based upon the 2007 EPA Fuel Economy Trends Report (EPA, 2007). Even though weight increased from about 3,000 pounds in 1987 to about 3,600 pounds by 2007, acceleration from zero to 60 miles per hour (mph) improved from 14.4 seconds in 1981 to less than 9.6 seconds by 2007. In addition, the proportion of manual transmissions, which are more fuel efficient than automatic transmissions, decreased from 30 percent in 1980 to less than 12 percent in 2007.

It is clear technology has been used for vehicle attributes that consumers demand or value more highly than fuel economy, such as performance, utility, luxury, and safety. The graph on the right of Fig. 5.10 compares the actual fuel economy for cars to what the fuel economy would have been if the technology were used solely for fuel economy instead of performance and other attributes. If the current car fleet were still at 1981 performance, weight, and transmission levels, passenger car corporate average fuel economy (CAFE) would be over 38 mpg instead of the current level of a little over 29 mpg. From 1987 to 2007, technology has gone into cars at a rate that could have improved fuel economy by about 1.4 percent per year, if it had not gone to other attributes more highly valued by the customer, such as performance, comfort, utility, and safety.

Figure 5.11 shows the same effects for light trucks. Weight has increased from about 3,700 to 4,700 pounds since 1987, the performance improvements have been virtually identical to that of cars, and use of manual transmissions has dropped from over 50 percent in 1981 to less than 4 percent. If light trucks were still at 1981

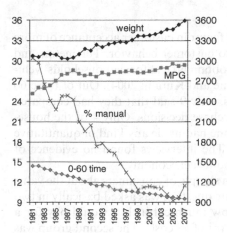
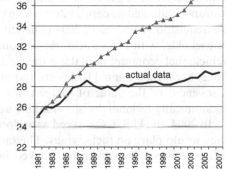

Fig. 5.10 Effect of attribute tradeoffs – cars

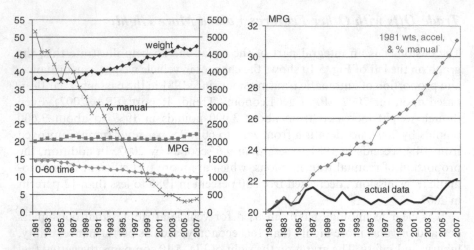

Fig. 5.11 Effect of attribute tradeoffs – light trucks

performance, weight, and transmission levels, LDT fuel economy would be about 31 mpg instead of just over 22 mpg. Since 1987, technology has gone into light trucks at a rate that could have improved fuel economy by about 1.6 percent per year.

There is no reason to believe that this rate of efficiency improvement will not continue into the foreseeable future. However, there is also no reason to believe that consumers will significantly change their purchase values, thereby eroding the energy and CO_2 emission savings from significant positive improvements in efficiency.

Customer Value of Fuel Savings

Customer value of fuel savings is a critical factor in customer acceptance of more expensive technology. To gain insight into customer behavior, Turrinetine and Kurani conducted in-depth interviews to obtain 60 California households' vehicle acquisition histories in 2004 (Turrentine and Kurani, 2004). Out of these 60 households, with 125 vehicle transactions, only 9 said that they had compared vehicle fuel economy in making their purchase decisions. Only 4 of the households knew their annual fuel costs and none had made any kind of quantitative assessment of the value of fuel savings. The interviews found no evidence of economically rational decision-making about fuel economy.

In 2004, the U.S. DOE asked a random sample of consumers about the value they would place on fuel savings. The same question was asked basically in two different ways. One group was asked how much extra they would pay for a vehicle that would save them $400 per year in fuel costs. The second group was asked how much annual fuel costs must be reduced to justify paying $1,200

Fig. 5.12 Consumer payback period – fuel savings

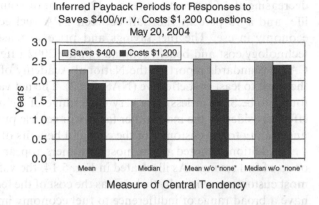

Inferred Payback Periods for Responses to
Saves $400/yr. v. Costs $1,200 Questions
May 20, 2004

more for a vehicle. Figure 5.12 shows the calculated mean and median payback
period for each group, as well as the results after eliminating respondents who
said they would not pay anything (Greene, 2004). The two groups gave generally consistent answers to the same question asked from two directions and
indicated a payback period ranging from 1.5 to 2.5 years. What really matters to
the consumer was the net value of their purchase decisions. Adding technology
to the vehicle can reduce fuel cost, but it also increases the price of the vehicle.

Figure 5.13 illustrates the net value to an "economically rational" consumer,
who values the full 14-year fuel savings (Greene, 2006). The analysis assumed a
$2.00 per gallon price for gasoline, 15,600 miles of initial driving per year,

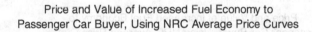

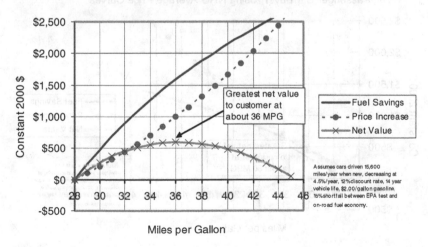

Fig. 5.13 Customer net value – 14 year payback

decreasing at 4.5 percent per year, a 12 percent discount rate, a 14-year vehicle life, and a 15 percent shortfall between EPA fuel economy and actual fuel economy in-use. The fuel savings and price increase were based upon the technology cost and benefit estimates in the 2002 Effectiveness and Impact of CAFE Standards report by the National Academy of Sciences, ordered from the most to least cost effective (NAS, 2002). The net value to the customer was found to be $500 or less for a very wide range of technology, which is a small effect considering the many other factors in vehicle purchase decisions and the uncertainty to the customer of the cost and benefits of the technology.

In addition, as noted above, most customers appear to value only three years or so of fuel savings. As illustrated in Fig. 5.14, the value of the fuel savings to most customers is virtually the same as the cost of the technology and, thus, they have a broad range of indifference to fuel economy improvements. Given this indifference, it is in the manufacturers' best interest to invest technology in offering better features, performance, utility, and safety, rather than more fuel economy.

Certainly, the recent increase in fuel prices will cause some change in customer behavior, but it is unlikely to cause major changes in customer demand. Also, there is no evidence that most customers would be willing to spend money just for the good of society. The view of most consumers is that it is the government's job to ensure that the needs of society are met, and even most environmentalists do not appear to be willing to pay extra themselves. Customers will generally accept cost increases if the government requires all other purchasers to also contribute to societal solutions.

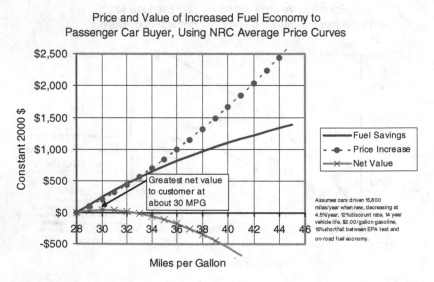

Fig. 5.14 Customer net value – 3 year payback

The Role of Government

Fuel prices influence vehicle purchase choices and reduce miles traveled. However, fuel price is not a good lever to pull technology into the fleet. The technology cost and fuel savings largely balance at current fuel prices and most customers greatly discount the fuel savings, resulting in little influence on highly complex and emotional purchase decisions.

The role of government is to step in when there is a gap between the values of consumers and society as a whole. This was the case several decades ago for vehicle pollutants that impacted human health and it is now the case for greenhouse gases affecting the climate. Climate change strategies will be more effective if they include government and customers, not just industry. The industry can provide a "pull" by providing products desired by the consumer and developing improved technology, but it cannot push customers into buying vehicles they do not want. Government programs to stimulate demand, provide incentives, and educate the customer can dramatically affect acceptance of new technologies and market penetration.

Given the rapid changes in technology, performance-based incentives and requirements are the best way to move the ball forward. Technology-specific mandates alone are not sufficient. In fact, previous attempts to mandate specific technologies have a poor track record, such as the attempt to promote methanol and the California electric vehicle mandate in the 1990s. If there are to be mandates, they should be stated in terms of performance requirements, with incentives and supported by research and development.

Leadtime and Costs

The low value of fuel savings by most customers is a barrier to technology implementation, but one that can be mitigated by properly designed federal requirements and/or incentives. There are clearly a large number of new technologies coming that will be cost effective and will enable substantial improvements in vehicle efficiency. The real constraint is how quickly these new technologies can be placed into production.

Most cost estimates assume, implicitly or explicitly, normal development, large volumes, and normal redesign cycles. However, the same cost estimates are frequently used for accelerated fuel economy requirements without consideration of leadtime. This is a substantial problem.

New Technologies = Huge Risks

The new vehicle market is very competitive. Vehicles of all types have steadily increased in quality, reliability, and safety. These features are highly valued by new vehicle customers, who have come to expect high levels of quality, reliability, and safety. This creates substantial risks for introduction of new technology.

At the same time, there are a multitude of new technologies available to reduce fuel consumption, with different synergies among technology packages and multiple ways to achieve the same efficiency improvements. Ironically, there are actually far too many technologies, which makes it impossible to see a clear path to the "best" solution. It is impossible to know how much costs will drop in the future with higher volume and further development. Estimates have been developed for this purpose, such as the estimates in the 2002 NAS CAFE Report (NAS, 2002). However, these estimates are only valid on average. The actual future cost reduction of an individual technology is highly variable.

Thus, one type of risk is that a manufacturer will be at a competitive disadvantage if the selected technology proves to be more expensive than other options. Complicating technology selection is the fact that the various technologies must compete against future advanced technology and engines, not just the engines now on the road. Thus, even the baseline for comparison is unknown. Development and deployment of advanced technologies must be an iterative process that continuously assesses changes in costs, benefits, and baseline. If proper leadtime for development and evaluation of multiple technologies is not allowed, manufacturers will be forced to guess at less than optimum solutions. Not only would this greatly increase costs to manufacturers and customers, but it could have grave competitive impacts.

Much worse is widespread adoption of a technology that does not meet the high customer expectations for performance and reliability. This would have a severe impact on the manufacturer's reputation and affect future sales and profitability. It can also set back acceptance of the technology for everyone. For example, there were a number of diesel engines rushed into production in the early 1980s, in response to consumer demand and large mandated fuel economy increases. These diesels proved to have severe reliability problems and set back customer acceptance of diesels in the United States for decades.

Leadtime Constraints

There are three steps to ensuring quality and reliability of new technology:

- Rigorous two to three year product development process, starting after the basic feasibility of the technology has been demonstrated.
- Proof in production on a limited number of vehicles for two to three years. This time is needed to assess the impact of higher volume and further development on costs and marketing impacts before committing to widespread deployment.
- Commercialization across the vehicle fleet in 5-year minimum product cycles.

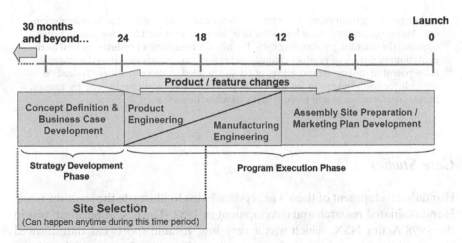

Fig. 5.15 Automotive product development timeline

This process assumes that the basic feasibility of the technology has already been demonstrated. Longer leadtime is needed for new technologies and for integrating multiple technologies, as the increased complexity exponentially increases the risks. Costs also increase dramatically if normal development cycles are not followed, as this greatly increases development costs, tooling costs, and the risk of mistakes.

The importance of leadtime was addressed in the 2002 NAS CAFE Report, as follows:

> Finding 15. Technology changes require very long lead times to be introduced into the manufacturers' product lines. Any policy that is implemented too aggressively in too short a period of time has the potential to adversely affect manufacturers, their suppliers, their employees, and consumers. Little can be done to improve the fuel economy of the new vehicle fleet for several years because production plans already are in place. The widespread penetration of even existing technologies will likely require 4 to 8 years. For emerging technologies that require additional research and development, this time lag can be considerably longer (NAS, 2002).

While the report's findings on technology costs and benefits have been widely used and quoted, the finding on leadtime has been largely ignored, even by the very organizations quoting the costs and benefits.

Step 1, the product development process, was assessed recently in a report by the Center for Automotive Research, "How Automakers Plan their Products." (CAR, 2007) Fig. 5.15 summarizes the product development timeline. The report describes in detail the factors manufacturers must assess in their development process and the leadtime constraints. For example, it states:

> Automobiles require long lead times for design, development and production planning (including tooling and supplier contracting). The process of developing a new program, whether for a new or redesigned vehicle or a powertrain, typically spans 2.5 years from concept to launch... because vehicle programs carry over a high level of components

and engineering from other programs, product changes are almost always evolution-ary. Moreover, intrinsic time lags—the two- to three-year lead time for product devel-opment, the even longer planning cycle for all of a company's products, as well as the evolutionary nature of product change—represent constraints that must be respected. Any potential policy requirements must acknowledge these realities. Indeed, it is difficult for automakers to do too much too fast. They are constrained by money, human resource issues and tooling costs, to name but a few.

Case Studies

Honda's development of the VTEC system helps to illustrate the leadtime issue. Honda initiated research and development in 1982. The first application was in the 1998 Acura NSX, which was a very low volume sports car that allowed Honda to gain experience before spreading the technology to high volume applications. Honda gradually added the technology across its product line until VTEC reached 100 percent penetration with the 2006 Honda Civic, roughly a quarter of a century after research was initiated and 18 years after the first market introduction.

EPA's fuel economy trends report includes industry-wide penetration of selected technologies, shown in Fig. 5.16 (EPA, 2007). The EPA did not collect information on the number of valves per cylinder before 2000 or on variable valve timing (VVT) before 1997, so the earlier penetration of these technologies is unknown. VVT includes cam-phasing systems, which are far simpler, but less effective than Honda's VTEC system.

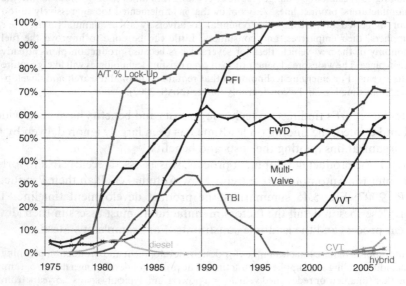

Fig. 5.16 Technology penetration rates

The throttle-body injection (TBI) curve shows what happens when a technology is displaced by a superior technology, in this case by port fuel injection, or PFI. This illustrates why technology choices need to be continuously assessed. The diesel trend is also a cautionary tale. Some manufacturers lost a great deal of investment in the diesel when fuel prices fell in the 1980s.

Only two technologies, PFI and torque converter lockup on automatic transmissions, A/T % Lock-Up in Fig. 5.15, show a sustained, reasonably rapid upward trend. PFI makes a particularly good case study, as it was a well known technology for many years before its use was driven by more stringent emission standards. Some manufacturers were already using PFI extensively when the EPA began gathering data in 1975 and PFI remained between about 3 and 6 percent of the fleet from 1975 to 1982. Thus, the technology was already proven and well understood when use began to increase in 1983. The use of PFI was driven primarily by a combination of new emission standards and a better understanding of catalyst operation. PFI has more precise air/fuel control and catalyst conversion efficiency is very sensitive to even minor changes in air/fuel ratio. Not only did PFI reduce emissions, therefore, but it did so while also allowing substantial reductions in the amount of precious metals in the catalyst, which largely paid for the additional cost of the PFI system. There were also secondary benefits with better cold start drivability and small improvements in fuel economy and performance. Yet, despite this favorable combination of a known technology, low cost, and multiple benefits, PFI still took 14 years to spread across the fleet.

Impacts of Aggressive Fuel Economy/Greenhouse Gas Requirements

Large annual increases in fuel economy require aggressive changes to every aspect of the vehicle. The industry does not have the resources to handle this level of change all at once. Even if it did, it would be too risky to implement the changes all at once.

Rapid technology deployment might be possible if it were known which technologies were going to perform best in vehicles over the long run, but this is unknown. Should aluminum, plastics, carbon fiber, or high strength steel be used to reduce weight? What are the impacts on safety? Can new tooling and assembly line methods be developed? Which materials are best for different parts? Which better aerodynamic shapes will be accepted by customers and which will lose market share? Which markets will accept downsized turbo-charged engines and which markets will require other solutions? Are central or side direct injection systems best? Is variable cylinder shutoff or variable valve timing the best solution and does this vary by application? Will variable valve timing technology be replaced by camless valves? Which transmission is best for each application? This is just a small sample of the questions that must be answered by manufacturers.

Aggressive fuel economy standards require a lot of choices be made in a short period of time. A company that rapidly implements advanced technologies and spreads them across its fleet is betting the choices are the right ones. If these choices turn out not to be optimal and other manufacturers develop solutions that work better at lower cost, the company could well go out of business.

Ironically, there are too many choices and too many potential technologies that have unknown final costs and benefits. Manufacturers need time to sort through them, test combinations, evaluate impacts on safety, performance, reliability, and drivability, see how costs drop with development and higher volumes, and see what the market will accept. This has to be done one step at a time.

A similar situation existed during the oil crisis in the 1970s. In response to fuel shortages, projections of rapidly escalating fuel prices, and CAFE standards, the combined car and light truck fleet fuel economy improved by 6.7 percent per year from 1976 to 1982. On the surface, this would seem to support aggressive fuel economy increases and refute the leadtime constraints. However, 40 percent of the fuel economy increase during this period was due to an 877 pound weight reduction in the entire fleet. This was the result of a one-time switch from rear to front wheel drive and redesign of extremely inefficient vehicle designs. Modern vehicles are already mostly front wheel drive and weight optimization, as much for performance as for fuel economy, has long been a priority, so the weight reduction realized from 1976 to 1982 cannot happen again.

This leaves a fuel economy improvement of 3.8 percent per year from 1976 to 1982 due to technology improvements, which is still an impressive number. However, a closer look at the results of this rapid makeover shows that unsafe and poor quality vehicles were rushed to market. The 2002 NAS Report concluded that the CAFE standards caused an extra 2,000–3,000 deaths annually (NAS, 2002). While these results should not be applied to future increases in fuel economy standards, there is little doubt that the rapid redesign of vehicles in the 1970s and early 1980s occurred without proper safety considerations and caused tens of thousands of deaths.

There are also many examples of poor quality vehicles and inadequate technologies rushed to market. The Chevy Chevette, Ford Pinto, and Chrysler K-cars all offered good fuel economy and sold well at the time, but developed reputations as relatively unreliable vehicles, damaging the reputations of the companies. General Motors introduced the V8-6-4 engine on most Cadillacs in 1981, but the technology was quickly retired due to a rash of unpredictable failures. The auto giant also rushed a V8 diesel into production, which proved to be underpowered and unreliable.

From a quality and customer acceptance point of view, it is fortunate that fuel prices dropped and fuel economy improvements deployed in response to CAFE mandates and fuel shortages largely stopped after 1982. Certainly the pause in CAFE increases has gone on far too long, but the pause did give the manufacturers time to correct the worst of their mistakes.

Some organizations have pointed to the rapid fuel economy improvements during the period from 1976 to 1982 as evidence that aggressive fuel economy increases are feasible. However, even a cursory look at what actually happened during this period should serve as a cautionary tale of what to expect if appropriate leadtime is not allowed. Certainly, the development of computer simulations since 1982 has enabled major improvements in the process of designing vehicles and more rapid design implementation. However, these improvements are largely needed to fulfill the much higher quality and safety expectations in today's market. With current customer expectations, a repeat of the 1976–1982 experience would be catastrophic.

Separate State and National Requirements Double Leadtime Constraints

Tailpipe pollutant emissions, such as hydrocarbons, carbon monoxide, and nitrogen oxides, are regulated at the federal level under the Clean Air Act (CAA). The CAA recognized California's unique air quality problems and their early leadership in regulating tailpipe emissions by allowing California to set their own vehicle emission standards, separate from federal requirements. California is the only state allowed to set different standards, although other states are allowed to adopt the California standards in place of federal standards.

California has recently extended their regulation of tailpipe emissions to include GHG emissions. However, control of GHGs is very different from most other pollutant emissions. Most tailpipe pollutant emissions primarily consist of unburned hydrocarbons, carbon monoxide emitted from partially burned gasoline, and nitrogen oxides created during the combustion process. Requirements to reduce these pollutant emissions have been met largely with catalysts and improved air/fuel control, which oxidize hydrocarbons and carbon monoxide into water and carbon dioxide and reduce nitrogen oxides into nitrogen and oxygen. In cases where California emission standards have been more stringent than federal standards, it was relatively simple to design a more sophisticated aftertreatment system for use in California vehicles. The system cost more, but, because it only involved a limited number of components, it was feasible and cost effective to design different systems for California and the rest of the nation.

The situation is fundamentally different with GHG control requirements. CO_2 is the end product of combustion, not a pollutant inadvertently created in small quantities during combustion. The average vehicle emits about 7 tons of CO_2 per year, which is far too much to trap. There is no easy way to convert CO_2 into another chemical. Except for switching to alternative fuels with lower carbon content, which is largely outside the control of the vehicle manufacturers, the only way to reduce CO_2 emissions is to reduce fuel use.

Thus, GHG control requirements are identical in their effects to CAFE requirements and require redesign of the entire vehicle and powertrain. Vehicles are designed, manufactured, marketed, and distributed nationally because of the huge development and tooling costs necessary to bring a vehicle to market. In fact, many vehicles are designed for and sold in multiple markets around the world, not just for the U.S. Engineering, tooling, and supplier resources simply do not exist to design and build separate fleets for different parts of the nation. Imposition of separate rules for states that adopt California GHG requirements would essentially double the leadtime constraints faced by the industry. Costs would explode and the rate at which technology could be transferred into the federal portion of the fleet would actually slow down, due to diversion of engineering resources and tooling capacity. These effects were recognized by the U.S. Congress when it preempted state regulation of fuel economy in the 1975 Energy Policy and Conservation Act and they remain valid today. The effects are also recognized by the rest of the world. No other country allows a state or province to regulate vehicle GHG emissions. In fact, the European Union does not even allow individual countries to regulate vehicle GHG emissions. The net result of state GHG requirements is vastly larger costs with little, if any, benefit over a federal program.

Leadtime in Japan and Europe

Japan and Europe have moved much more aggressively than the United States in the last two decades to improve vehicle efficiency and reduce GHG emissions. Fuel prices per gallon are $2.00–$5.00 higher than in the United States and both areas have implemented programs and consumer incentives to improve fuel economy.

Europe established a program to reduce vehicle carbon dioxide emissions from 185 grams per kilometer (g/km) in 1995 to 140 g/km in 2008. This translates into an annual fuel economy improvement rate of just 2.2 percent per year. Even so, Europe is not going to meet the 140 g/km target in 2008, despite its huge growth in diesel vehicle market share. Extension of the requirements is currently being debated. Japan recently established requirements to increase fuel economy from 13.6 km/l in 2005 to 16.8 km/l in 2016. The annual fuel economy improvement rate is 1.9 percent per year.

Overall, Japan and Europe are improving vehicle fuel economy by about 2 percent per year. This should raise serious concerns about the ability to increase fuel economy faster than 2 percent per year in the United States without impacting cost and risk, especially considering the low fuel price and lack of supporting customer incentives in the United States. It is also strong support for leadtime constraints.

Conclusions

There is a vast array of automotive technologies under development, all with varying degrees of promise and cost in the near and short term. Predicting which technologies will prove to be the best and when they will be ready is not possible. Manufacturers simply have to work on everything to ensure that they are not left behind.

One critical point is that improved conventional engines keep raising the bar regarding performance and efficiency. Major improvements in the internal combustion engine and important reductions in fuel consumption are likely to occur over the next several decades. This will reduce baseline fuel consumption from conventional vehicles and make it more difficult for alternative technologies and fuels to penetrate the market.

Hybrid electric technology is progressing rapidly, with costs decreasing, synergies with other technologies developing, and the potential for additional consumer features emerging. Emission control systems are coming that will enable diesels to meet the EPA Tier 2 standards. Both vehicle technologies should have steadily rising market shares in the short run, although appealing to different markets. Diesels will appeal more in rural market areas and for larger vehicles, while HEVs will appeal more in urban market areas and for smaller vehicles. In the long run, costs must greatly decline for mass market acceptance.

The ultimate goals are fuel cell or battery electric vehicles, but the timing is very unclear. Both have advantages and both require breakthroughs. PHEVs could prolong the fossil fuel era if a battery breakthrough occurs or if shortages of petroleum become prevalent.

While technology prospects are excellent, there are too many choices among potential technologies, all of which have unknown final costs and benefits. Manufacturers need time to sort through them, test combinations, evaluate impacts on safety, performance, reliability, and drivability, see how costs drop with development and higher volumes, and see what the market will accept and what it won't. This must be done one step at a time. If not, costs and risks escalate rapidly. Leadtime constraints have not been properly evaluated.

Technology cost estimates are usually applied without proper consideration of leadtime. Cost estimates are usually based upon normal development, large volumes, and normal redesign cycles. However, normal development and redesign cycles cannot handle annual fuel economy increases of more than about 2 percent per year. This is supported by the rate of fuel economy increases in Japan and Europe. More aggressive requirements require accelerated implementation of technology, with exponential increases in development costs, tooling costs, and risks of quality or safety problems. Certainly annual increases greater than 2 percent are feasible, but the rapid increases in cost and risk with larger annual fuel economy increases need to be assessed and balanced against the need of the nation to conserve energy and reduce greenhouse gas emissions. Instead of focusing on the costs and benefits of individual technologies, requirements should focus on the maximum rate of annual technology implementation and the incremental costs of increasing the implementation rate.

More aggressive requirements might be feasible if supported by measures targeted at consumer behavior. The real fuel cost of driving is still very low and most customers, rationally, value other attributes more highly than fuel economy. The industry can provide a "pull" by providing products desired by the consumer and developing improved technology, but customers cannot be pushed into buying vehicles they do not want. Government programs to stimulate demand, provide incentives, and educate the customer are needed to help acceptance of new technologies and market penetration. If there are to be mandates, they must be federal mandates and should be stated in terms of performance requirements, with incentives and supported by research and development.

References

Center for Automotive Research (CAR), How Automakers Plan Their Products, Primer on Automotive Business Planning, July 2007

Cheah, L, Evans, C, Bandivadekar, A, and Heywood, J, "Factor of Two: Halving the Fuel Consumption of New U.S. Automobiles by 2035", Massachusetts Institute of Technology, October 2007

Green, D, Duleep, K, and McManus, W, "Future Potential of Hybrid and Diesel Powertrains in the U.S. Light-Duty Vehicle Market", Oak Ridge National Laboratory, ORNL/TM-2004/181, August 2004

Greene, D L, "Why don't we just tax gasoline? Why we don't just tax gasoline," IAEE/USAEE Meetings, Washington, DC, July 10, 2004

Greene, D L, Presentation at Climate Change Policy Initiative, Washington, DC, October. 5, 2006.

Kliesch, J and Langer, T, "Plug-In Hybrids: An Environmental and Economic Performance Outlook", American Council for an Energy-Efficient Economy, Report # T061, September 2006

National Academy of Sciences (NAS), Effectiveness and Impact of CAFE Standards, Washington DC, 2002

Turrentine, T and Kurani, K, "Automobile Buyer Decisions about Fuel Economy and Fuel Efficiency", Institute of Transportation Studies, University of California-Davis, ITS-RR-04-31, September 2004

U.S. Environmental Protection Agency (EPA), Light-Duty Automotive Technology and Fuel Economy Trends: 1995 through 2007, EPA420-R-07-008, September 2007

Chapter 6
Heavy Duty Vehicle Fleet Technologies for Reducing Carbon Dioxide: An Industry Perspective

Anthony Greszler

Much of the focus on carbon dioxide (CO_2) reduction in the United States (U.S.) has been targeted at personal transport and passenger cars, despite the fact that over 20 percent of fuel consumed in United States surface transport is used in heavy commercial vehicles. The role of commercial vehicles is vital to the global economy and this segment is growing. Control of CO_2 emissions from heavy duty trucks requires unique metrics, technologies, and public policies.

Since the sale of Volvo's car division to Ford in 1999, AB Volvo has primarily concentrated on commercial vehicles through its work with Mack Trucks, Renault Trucks, Volvo Trucks, Nissan Diesel, Volvo Bus, and Volvo Construction Equipment. Volvo Powertrain is the primary supplier of engines, transmissions, and drivelines to these businesses. Volvo's environmental program focuses on reducing greenhouse gas (GHG) emissions, especially CO_2, from its truck engines.

The Role of Trucking in the U.S. Economy

While heavy duty commercial vehicles constitute a small percentage of the motor vehicle population, they consume a disproportionate amount of transport fuel, with most of this going to the largest, or class 8, trucks. The reasons for this are apparent. Long haul trucks spend a high percentage of their life on the road, hauling freight, which may be either very heavy or high in volume.

In fact, heavy duty ton-miles increased 55.5 percent from 1993 to 2002 (DOT, 2002), while vehicle miles increased 48 percent between 1990 and 2003 (FHA, 2004), reflecting the impact of increased freight and improved efficiency. Data for 2006, shown in Fig. 6.1, indicate 21 percent of surface transport fuel volume was burned in heavy trucks and buses, almost entirely in diesel engines. Another 13 percent went into offroad applications, such as construction and agricultural equipment, which share engine technologies with on-highway

A. Greszler
Volvo Powertrain Corp., 13302 Pennsylvania Ave., Hagerstown, MD 21742, USA

D. Sperling, J.S. Cannon (eds.), *Reducing Climate Impacts in the Transportation Sector*, DOI: 10.1007/978-1-4020-6979-6_6,
© Springer Science+Business Media B.V. 2009

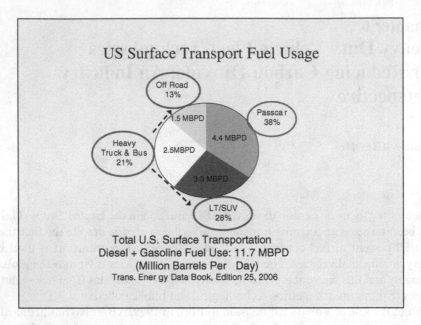

Fig. 6.1 U.S. surface transport fuel use

vehicles. It is also likely that the light truck segment, including sport utility vehicles (SUVs), will increase its use of diesel technology, as a result of increasing pressure to improve fuel efficiency and deploy "clean diesel" technology. If this occurs, diesel engines could become consumers of over 50 percent of U.S. surface transportation fuel.

U.S. heavy duty commercial vehicles are already a major consumer of petroleum fuels and the technologies related to this segment can influence an even bigger area. Despite this, there has been little government focus either to support research and development (R&D) or to develop informed public policy, except in the area of criteria pollutants, mainly particulate matter and nitrogen oxides. Moreover, heavy duty trucking requires a fuel with high mass energy density that is easily stored and transported. Hence, ground freight transport is particularly dependent on petroleum.

More than 80 percent of all communities in the United States are supplied with commercial goods exclusively by trucks. Trucks hauled 10.7 billion tons of freight in 2005, accounting for 69 percent of all freight by weight (ATA, 2007). Virtually every item sold to consumers traveled on a truck at some point on its way to the market. Typical domestically-manufactured product moves by truck an average of six times during production and distribution. Average imported products move four times by truck once reaching a domestic port. Trucking represents roughly 5 percent of the U.S. gross domestic product. The industry generated $625 billion in revenue during 2005, equivalent to 84 percent of all freight transportation revenues for all modes, including truck, air, water, rail and pipeline. Freight movement energy efficiency is lower for trucks than for rail or water, but trucking is often preferred or required because of its ability to delivery goods precisely from their point of origin to point of use.

Reducing CO_2 from Trucks

There are only two fundamental strategies for CO_2 reduction from heavy trucks: improved fuel efficiency and expanded use of alternative, low-carbon fuels. Using current or foreseeable technologies, it is unlikely that diesel engines will be replaced over the next 20 years, although it is likely that diesel hybrid systems will be deployed. Pure electric drivetrains, including battery, fuel cell or plug-in hybrid electric systems, will not be possible in the heavy duty truck sector without a major breakthrough in battery or fuel cell technology, except in very limited local operation. This is due to the high average power demand during normal highway operation, typically over 200 horsepower for class 8 trucks. Fortunately, the diesel engine can be adapted for a variety of fuels. There are also significant savings possible in freight movement efficiency, as discussed later in this chapter.

The commercial truck industry has multiple interests that are compatible with CO_2 reduction. Since the current fuel supply is almost entirely from fossil oil feedstock, the pending oil shortage is a significant concern. Price of crude oil and the resulting diesel price increases are a major part of the industry's operating cost. Also, with increasing public awareness of global warming, there is more focus by shippers and transporters on reducing CO_2 footprints.

Commercial trucking, especially long haul, where most fuel is consumed, places high importance on fuel efficiency, simply because fuel is a major part of a fleet's operating cost. In fact, at diesel prices of $3 per gallon, Fig. 6.2 shows that fuel may be the single biggest cost in many fleet operations, surpassing driver wages. A comparison of the initial purchase cost versus the cost of lifetime fuel consumed for cars and heavy trucks shows that the fuel cost is less than the purchase price for a car, while for a long-haul truck fuel cost is four times the vehicle purchase price. Thus, the fuel economy of long haul trucks is a critical factor in the purchase decision.

Even so, purchasers often do not select the best possible fuel economy vehicle for a variety of reasons. Because of the demand for on-time delivery and the high cost of breakdowns, truck owners are very conservative, leaning toward simplicity and proven reliability in purchase selections. With long haul annual driver turnover in excess of 100 percent, driver retention is also an important buying consideration. This places value on sleeper size and comfort, high power to allow sustained higher speeds since drivers are usually paid by the mile, and popular truck styling. Fuel saving features may interfere with optimum use of the truck or are subject to damage in certain road conditions.

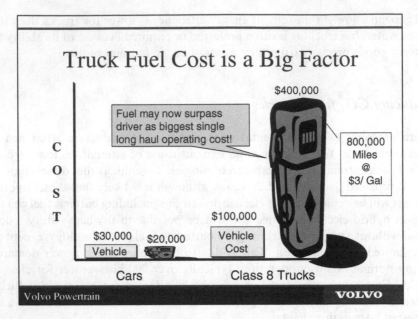

Fig. 6.2 Truck fuel cost is a big economic factor

Heavy Duty Vehicle Fuel Efficiency is a Complex Issue

The fuel efficiency for commercial vehicles is far more complex than miles per gallon (mpg). First, there are a wide variety of vehicles with different functions and duty cycles. For example, a long haul truck may cross the United States with long stretches at steady highway speed, while a garbage truck stops at every house and may never exceed 20 miles per hour (mph). In between these extremes are a vast range of vehicles including regional haul, local delivery, dump trucks and concrete mixers.

Even within the long-haul category, there exists a wide variety of trailers, trailer combinations, and loads. A vehicle optimized for a single application, as defined in a fuel economy test, could be poorly matched to its intended purpose, and actually deliver worse fuel economy. An example would be a tractor configured to pull an average van trailer but matched to a heavy load, forcing the driver to frequently downshift and use an engine speed range with poor efficiency.

An exaggerated example is shown in Fig. 6.3. If vehicle fuel economy, measured by mpg, is the key criterion for fuel efficiency, the pick-up truck easily comes out the winner. However, if ton-miles of cargo per gallon is the criterion, the road train is superior. If the criterion is selected to be cubic feet-miles per gallon, the superiority of a large vehicle is even greater. This is significant since the majority of trucks are volume limited rather than weight limited.

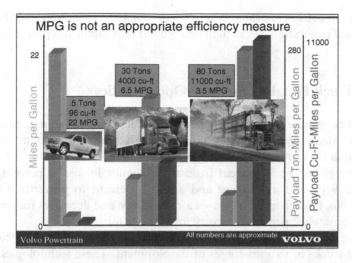

Fig. 6.3 MPG is not an appropriate efficiency measure

From the standpoint of goods movement, the key fuel economy criterion is freight movement efficiency, not vehicle mpg. It is also clear that freight movement efficiency can be heavily influenced by the allowable length, weight, trailer combinations and road congestion, as well as the base efficiency of the tractor-trailer combination. Indeed, Fig. 6.4 from a study by the International Council on Clean Transportation shows that modal energy intensity for truck

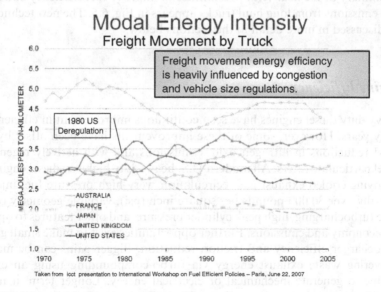

Fig. 6.4 Modal truck movement energy intensity

freight transport in Australia, where road-trains are permitted and there is a minimum of congestion, is less than half what it is in Japan and parts of Europe.

Heavy Truck Freight Efficiency Opportunities

Energy consumption in a truck is primarily due to air resistance, tire rolling resistance, auxiliary systems, and powertrain friction. The amount of energy consumed in each area varies with speed and load. At 65 mph with a full load, typical energy use is: 53 percent from aerodynamic losses; 32 percent rolling resistance; 9 percent auxiliaries; and about 5 percent to powertrain friction (DOE, 2006). Higher speed increases air resistance and decreases fuel economy by approximately 2 percent for every mph above 60.

There are a wide range of technologies to improve the fuel efficiency in long haul trucks in various stages of development. These technologies can be categorized as follows:

- Engine efficiency
- Transmission and driveline efficiency
- Hybridization
- Reduced rolling resistance
- Improved aerodynamics
- Weight reduction
- Reduced idling

A summary of the near term possibilities to improve fuel efficiency and reduce CO_2 emissions from long haul trucks appears in Fig. 6.5. The new technologies are discussed in more detail in the sections below.

Engine Efficiency

Heavy duty diesel engines have seen continuous improvement in efficiency for many years. However, some of these improvements have been offset by mandated reductions in nitrogen oxides emissions and with actively regenerated diesel particulate filters. All modern U.S. heavy duty highway diesel engines are deploying cooled exhaust gas recirculation, very high pressure fuel injection, typically over 30,000 pounds per square inch (psi), variable geometry or two-stage turbocharging, high peak cylinder pressure, and other features to optimize fuel economy and emissions. Further opportunities exist to make small gains in turbocharger efficiency and friction reduction. Bigger gains can be made by recovering waste exhaust energy via turbo-compounding using an exhaust turbine to generate mechanical or electrical energy. Longer term, it may be possible to use a bottoming cycle, such as a Sterling or Rankin cycle, or direct

Biggest Opportunities for Long Haul Trucks

Opportunity	Est. FE Gain	Technology Readiness	Issues/Obstacles
Low rolling resistance tires (super singles) on tractors and trailers	3%	Available for high volume use. Increasingly deployed.	Cost & life factors. Skepticism by operators. Trailer ownership split
Turbo Compound	3-5%	Concept proven with some production, but outside USA.	Cost and reliability Package space
Trailer side skirts	4%	Commercially available	Trailer/truck ratio >3 Trailer ownership split Skirt damage Knowledge/incentives
Mandatory Road Speed limit to 65 MPH (controlled via truck software)	5% average	Available in all class 8 trucks since mid 90's	Drivers paid by mile Car traffic meshing/safety Congressional Action
Eliminate Idling in sleeper mode	5-7%	Available: APU, battery, storage systems, shore power in some stops, engine stop-start systems, Idleaire system	Storage system performance Shore power availability IdleAire system availability & cost Cost & weight for on board systems California APU DPF requirement Stop/start cycle disturbs sleep
Increase weight, length, and trailer combination limits	Fewer trucks needed on road	None required	Safety concerns Road damage concerns State variations
Optimization of powertrain and engine to duty cycle	2-5%	Available	Customer awareness Adequate sales engineering support Variation in duty cycle
Trailer gap reduction	3%	Commercially Available. Deployed in some fleets.	Mix of trailers hauled. Turning radius reduction DPF size

Volvo Powertrain   **VOLVO**

Fig. 6.5 Opportunities to improve efficiency and reduce CO_2 emissions from long haul trucks

thermal/electric conversion, but these are not currently economical or practical within weight and space constraints in long haul trucks.

Other engine efficiency improvements can be achieved by further increases in peak cylinder pressure, more cooling of charge air and recirculated exhaust gases, better control of injection, and variable valve timing. However, these all add cost and complexity. Realistic gains of 3–5 percent efficiency can be expected in the next 5 years, while simultaneously reducing nitrogen oxides to near zero. Gains of up to 20 percent, from around 42 to 50 percent total efficiency, are being demonstrated in test conditions, but major costs, space constraints, and reliability issues must be overcome before these technologies can be deployed on commercial vehicles.

Transmission and Driveline Efficiency

Improved efficiency is also realized by deployment of advanced transmissions and by integration of the transmission with the engine. Fundamentally, these systems take gear shift management control from the driver to optimize fuel efficiency. Technology, now widely in use, deploys automatic shifting in the top few gears or reduced engine torque in lower gears to encourage drivers to use the top gears where best efficiency is achieved. The newest technology is the automated manual transmission (AMT), which allows gear shifting to be completely controlled by a computer using real or virtual sensors to determine the vehicle load, road grade and other factors to optimize gearing for any situation. AMTs use a manual gear box and clutch with actuators for shifting and clutching. The system retains the very high mechanical efficiency associated with direct gearing and avoids loses from torque converters typical of automobile transmissions.

Work is underway to develop transmissions capable of power shifting or continuously variable ratio. Either can eliminate turbo lag associated with power interruption during a gear shift, thereby allowing for a lower power engine to provide adequate acceleration. More significantly, these technologies can enable a narrow operating range for the diesel engine with potential for optimizing fuel economy. However, a continuously variable transmission typically has decreased transmission efficiency, reducing system gains.

Heavy Duty Hybrids

A variety of hybrid systems are currently in prototype, preproduction, or production stages. The most common is a parallel diesel electric configuration, although hydraulic and series electric are also deployed. A typical parallel electric hybrid system is shown in Fig. 6.6. The critical technology in need of development is energy storage. Due to very high energy associated with acceleration and deceleration of heavy duty vehicles, the mass and space for current

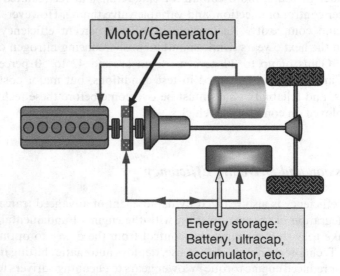

Motor/Generator

Energy storage:
Battery, ultracap,
accumulator, etc.

Fig. 6.6 Schematic of a parallel hybrid electric hybrid powertrain

energy storage technology limits effective energy recovery to relatively low speeds or small changes in speed in highway application.

Initial deployment of hybrid electric systems is targeted at stop-and-go duty cycles typical of urban operation, particularly in buses, refuse trucks, and local delivery and utility trucks. In these duty cycles, significant energy savings is gained by recovery of braking energy and reduction in engine idle time, with possible fuel savings of as much as 50 percent. However, total fuel consumption in theses applications is relatively small due to limited number of such vehicles, low speeds, and lower utilization.

The vast majority of U.S. diesel fuel is consumed in long-haul trucking where high speed, long distances, and heavy utilization are typical. For these applications, the benefits of braking energy recovery are on the order of 3 percent, less in flat terrain and more in rolling terrain. Under these conditions, use of hybrid electric technology can facilitate deployment of more efficient electric auxiliaries, such as fan drives, coolant pumps, air compressors, air conditioning, and power steering, as well as elimination of idling by using onboard energy storage and electric auxiliaries. It may also be possible to provide a low speed, creeper capability without running the engine during yard jockeying or in heavy traffic. Total system fuel savings of over 10 percent can be expected compared to a sleeper truck with no anti-idling technology.

Introduction of heavy hybrid electric trucks is still in a very early phase, utilizing expensive and low-volume or prototype systems. Costs can only come down with increased sales volume. Significant technology advancements are possible with adequate R&D. Government sales incentives and research funding are vital to moving ahead quickly.

Reduced Rolling Resistance

The simplest way to reduce rolling resistance is to assure proper tire inflation. Since tire pressure checks are often neglected, use of automatic tire inflation systems can save a significant amount of fuel. All heavy trucks use an onboard compressed air system for braking control and other functions, so a ready supply of compressed air is available. Of course, delivery of the air to rolling tires requires air tubing and a durable rotating seal at each wheel. Rolling resistance can also be significantly reduced by switching to low rolling resistance tires or "super single" tires. These tires use materials, construction, and tread designed to minimize tire heating and friction. Super singles are used to replace tandem tires with a single wide tire. Typical fuel savings of around 3 percent can be realized. The biggest hurdles to acceptance of these tires are concerns for tire life and safety issues or breakdowns due to loss of tire redundancy. Since almost all class 8 trucks also deploy tandem drive and trailer axles, there is already a second axle and tire to pick up load in the event of a blow-out. Industry acceptance is increasing with greater experience.

Improved Vehicle Aerodynamics

Since over half the energy expended to move a heavy truck is due to aerodynamic losses, a great deal of attention is given to this topic. However, much of the possible benefit requires trailer design changes and careful control of the combination vehicle because the great majority of heavy trucks are combination tractor-trailer rigs. Since trailer manufacture and purchase is outside the control of truck manufacturers, it is necessary to educate the trucking fleet operators to manage for fuel economy. Also, there are a wide variety of trailers, such as vans, flatbeds and tankers. Most attention has been focused on vans, since they comprise the majority of trailers.

For tractors, the key areas for improved aerodynamics are roof fairing, reduced frontal area, side skirts, and a front bumper air dam to reduce flow underneath the truck. Eliminating or shielding exposed components, such as air cleaners and exhaust pipes, are also important. A complete package to improve aerodynamics on a tractor can improve fuel economy by as much as 15 percent. However, many of these technologies are already widely deployed.

Another key area for aerodynamic improvement is in matching the trailer to the tractor. The objective is to create a smooth transition from tractor to trailer by matching the roof fairing to the trailer and to minimize the gap between the two. Since the gap is necessary to allow the vehicle to articulate for turning, gap extenders are deployed to minimize the air entry. Tractor to trailer matching can be accomplished in a good way for fleets dedicated to hauling standard vans. However, if a tractor is coupled with different trailers, it becomes much more difficult. For example, the fifth wheel may be moved back to allow flexibility in trailer choices, with increased gap. Roof fairings that match a standard van may have negative impact with a flatbed or tanker trailer, where a lower profile would be optimum.

Probably the biggest area for aerodynamic improvement is in the trailers. The two key technologies are trailer skirts and boat tails. The skirts enclose the wheels and side areas to prevent air entry and turbulence. Boat tails or vortex stabilizers are mounted behind the trailer to streamline the air flow over the rear of the trailer and reduce turbulence.

Fuel economy benefits of up to 10 percent can be realized through trailer aerodynamics. Although skirts are increasingly deployed, trailer aerodynamic retrofits are not in common use, due to several obstacles. Since trailers are often used as rolling warehouses, there are approximately three trailers in use for every tractor, effectively tripling the deployment cost. In addition, trailers ownership is often split between the shippers and the truck fleets, complicating cost allocation for retrofits. Even in use, skirts can be damaged by uneven surfaces or curbs on roads and in dock areas. Boat tails often interfere with loading and unloading operations. They also increase the trailer length, limiting cargo space due to state regulated length limits. There is a need for improved design, innovative incentives, length regulation allowances for boat tails, and recognition of trailer impact on fuel cost within shipping contracts in order to overcome these obstacles.

Weight Reduction

The benefits of weight reduction are more significant in weight-limited operations, typical of heavy material and tanker shipping, where less vehicle weight translates directly into increased freight weight, and improved freight movement efficiency. For volume-limited trucks, vehicle weight impacts energy input due to rolling resistance, acceleration, and hill climbing. Estimated impact on fuel economy for a class 8 truck is approximately 5 percent per 10,000 pounds of weight. Typical tare weight of a combination tractor trailer is approximately 30,000 pounds. Engineering development and material costs limit the practical potential for fuel savings from lower weight. Tractor weight and weight distribution typically results from the need to provide adequate frame stiffness, vehicle durability, maximum freight capacity, and weight distribution within axle weight constraints.

Idle Reduction

Most long haul trucks include a sleeper unit to allow drivers to maximize their time on the road and reduce hotel costs. These require "hotel functions," including heating, cooling and electricity supply for typical convenience appliances, such as onboard refrigerators, communications, and entertainment equipment. Traditionally, the trucks diesel engine has been left idling to power these systems, consuming from 0.8 to 1.0 gallons per hour. Idle time varies greatly depending on the route, temperature, and the driver arrangement. For a single driver in a long haul sleeper truck, regulations require a minimum 10 hour break per day, of which 8 must be in the sleeper berth. In this case, a 10 hour idling period is common. Maximum driving time for a single driver is 11 hours. At 10 gallons per hour while driving, reflecting a fuel economy of 6 mpg at 60 mph, the driving fuel consumption is 110 gallons. The idling fuel consumption is 9 gallons, or 0.9 gallons per hour, roughly 8 percent of the truck's total fuel use. This does not include idling at breaks and loading docks or for road congestion and stops.

There are a wide variety of devices available to reduce idle time, including engine start/stop systems based on cab temperature, diesel fired heaters, cold storage systems, battery powered cooling, plug-in electric systems, cold/hot air ducted in at truck stops, and diesel fired auxiliary power units. Mild hybrid electric systems can provide an optimal method for idle elimination, even during very short stops. All of these offer significant idle reduction potential, but also have drawbacks, such as added weight, limited operating time, availability, costs, and even potential air quality concerns from equipment not covered by stringent on-highway emissions standards. However, with the recent high cost of diesel fuel and increasing state anti-idling regulation, anti-idling systems are increasingly deployed.

Trucking Logistics

There are three categories of logistics management where significant fuel economy can be obtained. First is load management to maximize the time a truck operates at or near capacity. Although this is a very complex area, it is worth noting that new systems are available to track truck locations, to locate loads, to optimize the truck route and communicate to drivers. Second, it is possible to optimize vehicle routing to minimize distance traveled and avoid congestion for optimum delivery performance and fuel economy.

A third key area is vehicle management for fuel economy. Most trucks today already use systems such as road speed governors and gear down protection to encourage drivers to use highest possible gears. Newer systems are becoming available to optimize gearing under all operating conditions. It is also possible to use global positioning systems to adjust road speed limits, for example, to match local speed limits and road conditions, to anticipate grade and speed limit changes, or to reduce vehicle speed as a downhill or reduced speed area is approached.

Alternative Fuels

With annual freight growth averaging over 5 percent, the mitigation of GHG production from the transport sector will ultimately require the introduction of low carbon or carbon-free alternative fuels. Fortunately, the diesel engine is adaptable to a wide variety of fuels and engine manufacturers can adapt quickly to new fuel supplies, once it is clear what fuels will be available and when.

At this time, the primary alternative fuel available for use in diesel engines globally is biodiesel made mainly from soybeans in the United States and rapeseed in Europe. While there is considerable argument about the CO_2 savings from biodiesel, most estimates are in the range of 40–75 percent. Available land is limited for production of these grain crops and the yield is unlikely to exceed 5 percent of diesel consumption. Already, there is a big increase in price for grain crops used to produce biofuels, a shifting of land to this use, and concern about world food supply and cost.

Hence, it is essential to develop other nonfossil fuel alternatives. Key factors that must be considered in developing fuel alternatives initiatives include: sustainable resource availability, well-to-wheel energy efficiency and GHG impact, well-to-wheel emissions impact, cost, infrastructure requirements, energy density, fuel handling safety, health impacts, compatibility with existing vehicles, land use impacts, and public acceptance.

Volvo has studied the possibilities on the basis of well-to-wheel energy efficiency and GHG production. The results are shown in Fig. 6.7. An interesting finding is that dimethyl ether (DME) and methanol emerge as the best possible renewable fuels that can be made in large volume. DME has significant advantages in that it is non-toxic, burns without emitting particulate matter, can be liquefied at low pressure, like liquefied petroleum gas, and is an excellent high-cetane compression ignition fuel with very high engine efficiency potential. The biggest drawbacks for DME are that it lacks sufficient lubricity for use in a typical diesel fuel system and is an exceptionally strong solvent requiring careful material selection.

Most alternate fuel program in the United States are aimed at fuels which are compatible or nearly compatible with the existing fuel and vehicle infrastructure. From this perspective, synthetic diesel made from cellulosic biomass is a strong candidate. Almost any biomass, including waste products, wood, plant residue, or biomass fuel crops like switchgrass, can be gasified and converted to synthetic diesel, a pure hydrocarbon that can be mixed with petroleum diesel, shipped via pipelines and used in any modern diesel engine with no problems. It also has little or no sulfur, a high cetane number, and no aromatic compounds, so it burns cleaner than petroleum diesel. Current processes for synthetic diesel production are expensive, but rapidly becoming competitive as technology

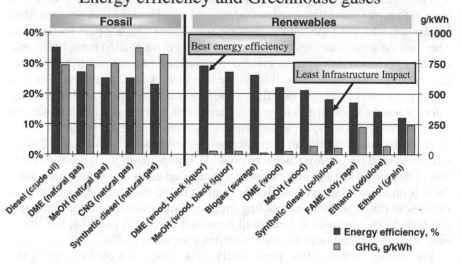

Fig. 6.7 "Well-to-Wheel" analysis of fuels for heavy duty trucks

improves and petroleum cost escalates. However, even as these processes become competitive with fossil oil, appropriate policies are needed until commercial application is fully proven. It may also be necessary to assure minimum price support if petroleum cost drops.

Public Policy

Freight movement efficiency is heavily influenced by vehicle size and weight regulations, speed limits, and congestion. All these are within the domain of public policy. In the United States, weight and size limits are controlled by individual states. These limits vary significantly, with some states allowing longer or heavier trailers, some allowing triple trailers, and variations in axle weight limits. A long haul trucker needs to load for the most stringent state on the planned route. Consistent rules allowing the longest and heaviest possible trucks and combinations can greatly improve the freight movement efficiency by minimizing the number of trucks required.

Probably the simplest and easiest public policy to improve fuel efficiency in the United States is to set mandatory road speed limits for trucks. This has already been done in most other countries. Since fuel economy decreases by approximately 2 percent per mph above 60, even a small reduction in average speed would have a big impact. Virtually all trucks built since 1995 have road speed governors that can be set by fleet managers to limit maximum vehicle speed. A maximum governed speed limit could be applied to these existing trucks and be preset on new trucks at the factory where the truck is originally manufactured.

The American Trucking Association has requested a mandatory road speed limit of 65 mph to save fuel, improve safety, and avoid competition between fleets for drivers, who want the highest possible speed since they are paid by the mile. There are no good estimates for how many trucks drive over 65 mph, but even a casual observation indicates the number is significant in some areas. A 5 percent fuel savings could result from this simple measure. A similar measure for cars could produce even bigger benefits, while matching car speed with trucks for safety. Another alternative is to provide "truck only" lanes on major highways to allow trucks to run at slower speed without interfering with car traffic.

Highway congestion is an ever increasing problem for all vehicles. Trucks will avoid such congestion whenever possible, but increasingly, it is not possible due to delivery schedules and expanding congested areas and time spans. Since there is no alternative to truck freight in most cities, the optimal solution is to reduce car traffic through carpooling, public transit, housing pattern management, and other measures. New road construction is also possible, but this is only a temporary measure because it invites even more traffic.

For certain commodities, particularly bulk goods like coal or grain, rail freight is significantly more efficient than trucks. For lighter goods, there is a wide range of estimates comparing the two modes. From a public policy standpoint, it makes sense to use rail freight as much as possible for heavy bulk goods and to use intermodal transport where it is viable and an energy efficient alternative. However, the U.S. rail infrastructure is already strained and also needs investment.

Recently, there is much discussion at federal and state levels for heavy truck fuel economy regulation, along the lines of the corporate average fuel economy program that currently applies to cars and light trucks. Such a simple approach can lead to negative effects if it results in smaller trucks or trucks improperly set up for the loads they need to haul in order to meet a fuel economy standard that may not be representative of the intended use. This can be addressed partially by establishing a variety of duty cycles, but will create a complex regulation.

Another major complication for heavy truck fuel economy regulation is that trailers are not manufactured, controlled, or sold by truck manufacturers. A proper match between tractor and trailer is essential for good fuel efficiency. This can only be controlled by truck fleets and drivers.

Finally, there are vehicle designs and logistics technologies that are unlikely to be verified in a standard vehicle test cycle. A careful study of these issues should be carried out before establishing GHG regulations for heavy duty trucks. Given the complexity, it may be more appropriate to use fuel or carbon taxes as a method to increase the market value of fuel efficient freight transport.

Summary and Conclusions

The mitigation of GHG impacts from heavy truck transportation will require a complex and evolving set of technologies. There are many claims of technology possibilities that could lead to huge efficiency improvement gains of over 50 percent with minimal effort. Many of these are unrealistic or double count interacting effects. However, it should be possible to achieve 20–30 percent efficiency improvement if proven technologies are fully applied to tractors and trailers. This will require education and incentives to truck owners along with well-designed regulations. Both tractor and trailer must be included and properly matched. Changes in truck size and weight regulations should be made to accommodate weight and length increases associated with fuel saving devices, so there is no penalty in lost freight capacity.

If freight volume continues to increase at over 5 percent annually, the only realistic way to obtain significant GHG reductions is to deploy low carbon alternative fuels. Alternative fuel technologies exist, but need further development and their cost must be reduced.

References

American Trucking Association (ATA), American Trucking Trends, 2005–2006, 2007
Federal Highway Administration (FHA), Highway Statistics 2003, Washington, DC, 2004
U.S. Department of Energy (DOE), Energy Efficiency and Renewable Energy, 21st Century
    Truck Partnership, 21CTP-003, December 2006
U.S. Department of Transportation (DOT), Bureau of Transportation Statistics and U.S.
    Department of Commerce, U.S. Census Bureau, 2002 Economic Census, Transportation
    2002 Commodity Flow Survey, Washington, DC, 2002

Chapter 7
Beyond Congestion: Transportation's Role in Managing VMT for Climate Outcomes

David G. Burwell

Public concern over the rapidity of climate changes, and the potentially catastrophic consequences of such changes to economic, social, and biological systems has exploded over the last two years (IPCC, 2007). At the same time, traffic congestion has continued to worsen, and now costs urban travelers an average of 38 hours in travel delay annually (TTI, 2007). This chapter addresses the question whether or not reducing vehicle miles traveled (VMT) is a sensible strategy for reducing both traffic congestion and transportation-related emissions of carbon dioxide (CO_2), the primary greenhouse gas contributor from the transportation sector.

The answer is yes, but such efforts are presently being lead by state government Departments of Transportation (DOTs) interested in congestion reduction, not climate, and the focus of such DOT efforts is convincing local governments to pay attention to the traffic generation implications of their land use decisions, not on DOT initiatives to reduce the VMT implications of their own actions (Toth, 2007). While, in the past, the core mission of transportation managers has been to meet the mobility needs of a growing economy, this mission has more recently been restated to focus more precisely on congestion relief. Meanwhile transportation-related CO_2 has been perceived as a concern best addressed by vehicle manufacturers, fuel providers and land use planners, not the owners and managers of the transportation system itself.

This reasoning is gradually changing due to the increased recognition of the mutuality of interest in managing transportation systems for both congestion and climate outcomes. The subtle mission-shift from mobility (which views VMT as an unqualified benefit that adds to consumer utility and comfort) to congestion relief (which is more ambiguous about VMT growth when it reduces overall system efficiency) allows transportation managers and climate advocates to coalesce around VMT reduction as a common strategy for both congestion relief and climate protection. When these two outcomes are addressed together, strategies and partnerships with non-transportation agencies, interest

D.G. Burwell
BBG Group, 7008 Rainswood Ct., Bethesda, Md. 20817, USA

D. Sperling, J.S. Cannon (eds.), *Reducing Climate Impacts in the Transportation Sector*, DOI: 10.1007/978-1-4020-6979-6_7,
© Springer Science+Business Media B.V. 2009

groups, and private sector partners expand. Also, managing VMT coopera-
tively across disciplines appears to be more effective than single-discipline
efforts. The potential exists for such cooperative efforts to make meaningful
reductions in VMT growth and to remove the barriers transportation system
managers face in achieving such reductions.

VMT Trends: Setting The Goal

Over the 45-year period from 1950 to 1995, VMT in the United States (U.S.)
grew at an average annual rate of 3.5 percent. Growth then moderated between
1995 and 2005 to a slower, but still robust annual rate of 2.1 percent (ORNL,
2007; FHWA, 2005). During the same period, the U.S. population increased at
a much slower rate, an average of 1.2 percent from 1950 to 1995 and 1.1 percent
from 1995 to 2005). As a result, while population increased 96 percent over the
55-year period between 1950 and 2005, from 151.3 million to 299 million, VMT
increased more than 6500 percent, from about 43 billion miles annually in 1950
to 2.99 trillion miles in 2005. Looking forward, over the next 22 years VMT is
projected to continue to increase at an average annual rate of 1.6 percent, which
is the baseline used for projections of transportation energy use through 2030.
(DOE/EIA) Over the same period. the U.S. population is expected to grow at
about 1 percent annually. Thus, VMT growth is expected to continue to out-
pace population growth.

The continued variance between population and VMT growth rates leads to
an inevitable result: more people driving longer distances, mostly alone.
Between 1950 and 1995, VMT per capita increased from 3,029 miles per year
to 9,098 miles per year, a 1.9 percent increase per year, and increased further to
10,087 miles per year in 2005, reflecting a more moderate 1.0 percent increase
per year (ORNL, 2007; FHWA, 2005). At the present growth rate, VMT per
capita will be about 15,550 miles per year by 2055.

Climate scientists now estimate that global carbon emissions must stabilize
at about 70–80 percent below 1990 levels of CO_2 equivalent emissions (CO_{2e}) to
keep global warming contained at about 2 degrees Centigrade above historic
levels. Warming above this level stimulates unacceptable consequences, including the probable loss of over one-third of global species, sea level rise that
displaces hundreds of millions of people, and increased disease and famine
(IPCC, 2007). U.S. transportation CO_{2e} emissions in 2006 were 26 percent
above 1990 emissions. At a 1.9 percent annual VMT growth rate, VMT will
increase from 2.99 trillion miles in 2005 to about seven trillion in 2055, or about
2.34 times today's level (AASHTO, 2007a). Absent additional improvements in
fuel and vehicle efficiency that would put 2005 transportation CO_{2e} emissions
from surface transportation in 2055 293 percent above 1990 levels.

If VMT grows as projected by AASHTO, transportation-related CO_2
emissions per mile traveled, called the CO_2 intensity, must decrease by over

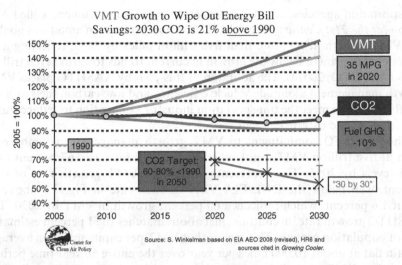

Fig. 7.1 Transporation CO_2 Reductions Under 2007 Energy Act v. CO_2 targets

90 percent per vehicle average through vehicle and fuel efficiencies by 2055 to achieve reductions to levels 70 percent below 1990 emissions. Since a large percentage of the vehicle fleet in 2055 will not have achieved 90 percent reductions in combined vehicle and fuel CO_2 intensity, the remainder of the fleet (newer vehicles) must sufficiently exceed the 90 percent goal to offset the shortfall. Absent a ban on the internal combustion engine, this is virtually impossible.

Even reducing transportation CO_2 to 1990 levels by relying solely on vehicle and fuel technology improvements will be difficult. The recently-enacted Energy Independence and Security Act of 2007 now mandates a 40 percent improvement in vehicle efficiency by 2020 and provides significant incentives for production of low-carbon fuels.[1] These measures, if fully implemented in a timely manner, approximate on a national basis the vehicle efficiency and low-carbon fuels requirements established under California law AB 32. However, as shown in the Fig. 7.1 above, assuming the targets set by AB 32 and the Energy Independence Act are both met nationwide, U.S. vehicle-related carbon emissions will still be 21 percent above 1990 levels (2 percent below 2005 levels) due to continued increases in vehicle miles traveled. This compares to a global target of a 20 percent reduction *below* 1990 carbon levels by 2020 and 70 percent below 1990 carbon levels by 2050.

Transportation industry leaders have acknowledged that improved vehicle efficiency, combined with a rapid conversion to low-carbon fuels, will be insufficient to meet transportation CO_2 reduction targets. In June 2007, the Board of Directors of the American Association of State Highway and Transportation Officials (AASHTO), the national trade association for state

---

[1] Signed into law December 19, 2007.

transportation agencies, adopted a new strategic vision document, called *New Vision for the 21st Century*. The AASHTO report sets out an ambitious goal to cap VMT growth at no more than five trillion miles by 2055, reflecting a 50 percent cut in growth below the growth in current trends towards seven trillion miles (AASHTO, 2007a). The adoption of this goal by AASHTO places VMT growth management alongside vehicle efficiency and low-carbon fuels as a co-equal strategy to meet the transportation industry's obligation to reduce transportation-related carbon emissions.

The AASHTO commitment to VMT growth management is ambitious. Even at five trillion VMT, total vehicle travel still increases 67 percent over 2005 levels by 2055. This represents an annual VMT growth rate of 0.95 percent, well below the DOE/EIA VMT growth rate in Fig. 7.1 above of about 1.6 percent, which projects a 60 percent growth in VMT by 2030. The AASHTO growth rate, in contrast, just about matches the 1 percent estimated rate of population growth. This means that VMT per capita must, on average, remain flat at about 10,000 miles per year over the entire 50 year time period. Given that VMT growth per capita is now about 1 percent annually, any further increase in VMT per capita must be overset by negative VMT growth per capita in the out years. This can only happen if distances between origins and destination, both freight and passenger, shrink, and trips choices increase, with more people using alternative modes of travel. This has not happened since World War II.

Still, it is significant that AASHTO, which represents transportation system owners and managers, has adopted this VMT performance metric. The traditional view is that, as the economy grows, people will drive more. This is no longer accepted wisdom. By adopting a long-term VMT cap of no more than five trillion VMT by 2055, AASHTO is accepting the challenge of meeting society's access needs for a growing economy, while keeping vehicle miles per capita growth flat. This is a significant shift, consistent with a new focus on congestion relief through VMT growth reductions rather than a sole reliance on new capacity as a congestion relief strategy.

VMT Measurement: State of The Practice

What gets measured, gets managed. If capping VMT at no more than five trillion miles over 50 years is the goal, reducing VMT per capita is the performance metric for tracking progress towards this goal. Since growth of VMT at a faster rate than population growth in the early years must be offset by reductions in VMT per capita in the out years, it is important to accurately measure and track VMT on an ongoing basis. This includes an understanding of:

- VMT measurement methodologies
- Demographic trends affecting VMT growth
- State and regional VMT and VMT per capita trends.

This requires extensive collaboration between state DOTs and other state agencies and policymakers to make accurate VMT measurement a public priority, and to assign roles, responsibilities and accountability for VMT management.

Unfortunately, the state of VMT measurement today is based mostly on guesswork. There are two basic methodologies and reporting systems. The first is by counting traffic. States conduct traffic counts on a small sample of the state road system, and an even smaller portion of the local road system, then extrapolate the results to the entire system through estimates. These estimates are then reported to the Federal Highway Administration (FHWA), updated annually through the FHWA's Highway Performance Management System (HPMS), and published in *Highway Statistics*. The results are used to measure and track VMT by road type and function.

A different VMT measurement methodology is used for purposes of apportioning federal transportation dollars to states. Distribution of federal transportation assistance to states is determined, in part, by statewide VMT expressed as a percentage of nationwide VMT. This percentage is based on variables that include the total statewide vehicle registrations, the average age of vehicles by vehicle class, and statewide gasoline tax receipts. These factors are then analyzed to make a best guess calculation of total statewide VMT. No traffic counts are conducted.

These methodologies are of limited value to measuring and managing VMT. First, state road counts are conducted on an average of every three years. Not only are statewide VMT totals calculated from a small sample, the results are updated through trend analysis, not counts. Second, the extent to which a sampling of VMT use on a few road segments, even with similar geographies and functional classifications, represents use throughout the entire system is uncertain. Third, some counts are not updated for more than three years. Finally, a handful of states have stopped providing updated counts to FHWA entirely, in which case previously submitted data is simply updated by trend analysis. The combination of small samples, irregular updates, the use of ambiguously relevant road segments, and further annual updates using trend analysis, makes it almost impossible to determine statewide or regional VMT based on these counts.

VMT analysis based on vehicle registrations, fleet age and gasoline sales are also of limited value. The only variable in this calculation that would reflect actual vehicle use is gasoline sales, since vehicle registrations and vehicle age don't, by themselves, reflect level of use. Gasoline sales reflect level of use, but not vehicle fuel efficiency, driving habits, or cross-regional or cross-state travel. They also do not reflect road conditions, such as level of congestion, which directly affects on-the-road fuel efficiency. This calculation also provides little information on VMT by region, and no information on VMT by road type. This is not an adequate dataset to measure and manage VMT, or to understand how policy intervention changes driving habits.

If VMT is going to be both a system performance metric and a way to measure the effect of policy intervention, present VMT measurement tools must be improved and new measurement strategies adopted. New technologies that allow vehicle mileage to be tracked on a real-time, open-road basis are presently being tested in Oregon as a potential strategy for tying transportation finance to the actual amount of road capacity vehicles consume. Connecting finance to miles traveled rather than fuel burned is known as a VMT tax, which may well become a favored methodology for financing transportation projects in the future. It has the tentative support of AASHTO and other leaders of the transportation sector (AASHTO, 2007b: Revenue) If and when transportation finance policy includes a VMT tax, accurate VMT measurement and tracking will be greatly improved.

Demographic Trends Affecting VMT

VMT at the national level continues to grow at about 2 percent annually, and VMT per capita at about 1 percent annually. This rate is not uniform across regions. In 2002, for example, VMT in the Puget Sound area in Washington grew at a rate of 1.3 percent, below both the population growth rate of 1.4 percent and the employment growth rate of 1.5 percent. These data indicate that economic activity in the region is not dependent on ever-increasing levels of VMT (Puget Sound Regional Council, 2007). The Portland, Oregon, portion of Multnomah County achieved a remarkable 7.5 percent reduction in VMT per capita in the period from 1996 to 2006, primarily through coordinated transportation and land use planning, and targeted investments to increase transportation choices (Burkholder, 2007). The Los Angeles and San Francisco Bay areas in California have reported small decreases in VMT per capita, again through growth management and investment in travel choices, while statewide VMT per capita has remained flat since 2000 at 24.6 miles per day (MTC, 2007). While these examples of flat or declining VMT per capita growth are anecdotal, they are worth studying in greater detail to determine how they were achieved during a period of robust economic growth.

Some demographic realities hold promise for further decreasing the rate of VMT growth, even absent policy intervention. By identifying such changes, policies may be developed to further support these demographic trends. The following examples suggest that a 50-year, 50 percent average reduction in VMT growth will not be easy, but it is possible.

The Aging of the Baby Boomers

During the period from 2011 to 2030, roughly 51 million of the 78 million "baby boomers" born between 1946 and 1964 will pass 65 years in age (Brookings, 2007). It is logical to assume that, as they do, their driving will begin to decline. The data support this conclusion, showing about a 30 percent drop-off in VMT by men in the 65–74 age period as compared to pre-65 age groups, with further rapid declines after age 75. For women, the decline is less marked, about 20 percent (Hu, 2000). The difference between male and female VMT reductions is assumed to be that more men retire at age 65 and their commute-related VMT drops faster. On the other hand, the same study shows an overall increase in VMT within the 65–74 age group compared to previous generations of seniors. The increase is about 35 percent. Therefore, Baby Boomers will probably drive more than previous generations of seniors, although they will drive less than when they were younger. On balance, it appears that the VMT trend from the aging of this generational cohort will be downward. This is not to imply that the overall population will not grow. It will, at an average rate of 30 million more residents per decade. However, since this age cohort will remain a larger-than-average percentage of the overall population, as it stops driving, VMT per capita, on average, goes down.

The Urbanization of America

In 2005, the split between urban and rural VMT was 1.952 trillion miles urban, 65.5 percent of the total VMT, and 1.038 trillion miles rural, or 34.5 percent of the total. This is roughly 2:1 urban to rural split. Average annual household VMT for urban areas is 19,300 miles while it is 28,400 annual VMT for rural areas. However, average daily traffic (ADT) on rural interstates has been declining since 2002, from about 260 percent above 1970 levels to about 225 percent above 1970 levels (FHWA, 2005). The implication is that, as the United States becomes more urbanized, household VMT declines, partly because destinations are closer together and partly because urbanized areas tend to offer more mode choices. As observed by the Center for Metropolitan Policy, part of the Brookings Institution, the United States is no longer defined by the Jeffersonian, small-town, agriculture-based community. It is becoming a "Metropolitan Nation." This will have a dampening effect on VMT growth (Brookings, 2007).

Decline in the Rate of Road Construction

The significant expansion of U.S. road capacity during the last half of the 20th Century, stimulated and financed by the federal Interstate and Defense Highway Assistance Act of 1956, provided access to jobs and opportunities to millions of Americans. When the Interstate System was finally declared complete in the Intermodal Surface Transportation Efficiency Act of 1991, cumulative federal expenditures on Interstate System construction had exceeded $135 billion. These expenditures were part of a strategy for reducing congestion. It is now generally accepted among transportation experts, however, that new road capacity has a stimulating effect on VMT due to "latent" or "induced" demand for access to such capacity, especially if it is not priced (ENO, 2002; Noland and

Cowart, 2000). However, since Congress declared the Interstate System complete in 1991, transportation policy has shifted to favoring management solutions rather than focusing on new construction as the primary strategy for dealing with congestion. Several states, such as New Jersey, Massachusetts, Oregon and parts of California, have adopted "fix it first" policies directing that the existing system be in a state of good repair before building new road capacity.

As state DOTs focus an increasing percentage of their human and financial resources on managing the existing road system, VMT growth resulting from induced demand will decline. This will have a dampening effect on VMT growth and already appears to have done so given the relatively sharp decline in VMT growth from 3.5 percent annually in the 1950–1995 period to a more moderate 2.1 percent over the 1995–2005 period.

Workforce Saturation

Another historical stimulant to VMT has been the addition to the workforce of women, minorities, and non-U.S. residents who, whether due to discriminatory employment policies, strict immigration laws, or a culture that discouraged women from working outside the home, were previous excluded. The entry of these new cohorts into the workforce over the last 50 years resulted in more people traveling more often. The rate of entry into the workforce of these previously excluded workers is now declining. This has a dampening effect on VMT growth.

These are only some of the more obvious demographic changes affecting the rate of VMT growth. They do not affect the overall direction of VMT which, due to an ever-growing population base, will continue to climb. However, there is anecdotal evidence that VMT growth is leveling off, as noted, in the following Fig. 7.2 prepared by the Center for Clean Air Policy based on USDOT, USEPA, USDOE and USDOC data.

The point of this discussion is not to promote a sense of complacency. Guiding the VMT growth rate lower will be a very hard job. It will require transportation system managers, working closely with their community and sister agency partners, to maintain a strong, persistent focus on this goal over a long period of time. The essential point here is that the past is not the future, and trend is not destiny. It can be done.

The Effect of Transportation Policy Intervention on VMT

The demographic changes discussed above make VMT growth reduction possible, but not inevitable. Policy intervention will be required. Travelers will have to change their travel behavior. Communities must become more compact. Average network speed must decline without increasing average trip time or

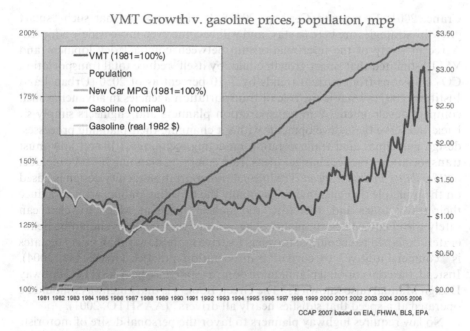

Fig. 7.2 VMT growth and Gasoline prices, population and fuel economy

reducing reliability. Access to congested road space, at least during peak hours when demand for that space is at its peak, must be priced. Trip choices must increase. This will require a lot of creativity, ingenuity, and a rethinking of past assumptions about best agency practice. Maximizing motorist convenience and efficiency no matter what will have to cede priority to a more balanced, nuanced and negotiated set of outcomes.

Transportation agencies are just now beginning to incorporate VMT reduction strategies into their internal planning processes. Congestion relief is still the primary objective of such efforts and funding shortfalls are the primary motivation. "To reduce traffic maybe we have to reduce the number of trips on the highway," notes John Lettiere, former DOT commissioner in New Jersey (Swope, 2005). However, the mechanism for achieving this objective is almost exclusively to encourage local governments to stop exporting traffic generated by new developments onto the state highway system. Only a few state DOTs are affirmatively investing transportation dollars into VMT reduction strategies. Some ideas for such reduction from already-adopted best transportation practice follow.

Land Use Planning and System Design Strategies

It is now well established that compact, mixed-use developments tend to generate less VMT/capita than sprawling, single-use developments (Boarnet and

Crane, 2001; Ewing and Cervero, 2001). It also appears that such "smart growth" strategies are here to stay, and will become even more widely adopted. A recent study of the inter-relationship between compact development and VMT estimated that smart growth could, by itself, reduce total transportation CO_2 emissions from current trends by 7–10 percent as of 2050 (Urban Land Institute, 2007). What is the role of transportation agencies in advancing such compact development? Can transportation planners and engineers simply sit back and serve these developments without changing their planning processes, design guidelines, and transportation modeling techniques? If not, how must transportation practice change? Ideas from emerging planning practice include:

Flexibility in design speed: Traditional transportation facility design is based on the principle that customers, essentially the traveling public, want to reduce their travel times and, therefore, the highest speed at which travelers can safely travel equals design quality. The concept that facilities should be deliberately designed to encourage travelers to drive more slowly than safety dictates is "counter-intuitive" to current transportation practice (AASHTO, 2004). Instead, the conventional planning wisdom, as stated in the AASHTO Highway Design Manual (also known as "The Green Book") is that "highways should be operated at a speed that satisfies nearly all drivers" (AASHTO, 2001).

No law requires highway planners to favor the personal desire of motorists to travel as fast as safely possible over community and public values. Yet, as new vehicle technology and safety features allow car manufacturers to build cars that can be handled safely at higher and higher speeds and market this "high performance" feature, highway design accommodates higher speeds. As speed increases, distances between destinations explode. Sprawl and more VMT are the results. Smart growth planners cannot achieve the objective of creating networks of compact, mixed-use communities when both car manufacturers and highway planners have their design foot on the accelerator.

Smart growth planning and highway planning must become much more closely aligned to achieve meaningful VMT reductions. This means that entire road networks, not just individual streets or subdivisions, must be designed for slower travel speeds. Public policy also plays a role—especially public policy favoring climate protection and energy efficiency. Slower, denser road networks support walkable, energy-efficient, and climate-friendly communities.

An important caveat is that slow road networks require the existence of fast road networks. Just as not all highways and streets need to designed to "satisfy nearly all drivers," neither should all highways be designed to satisfy nearly all walkers. Design speeds that are appropriate for "the social world" (city centers, neighborhoods, gathering places etc.), are not appropriate for "the traffic world" (inter-urban passenger and freight travel corridors) (Engwicht, 2005) (Fig. 7.3). High-speed systems, such as the Interstate System and its connectors, are needed to serve intraregional and interregional passenger and freight needs. These high-speed road networks should be designed to focus on their primary mission to facilitate long-distance travel. This means access controls. When managers of high-speed roads allow access through curb cuts to serve

Traffic World	Social World
Uniform	Diverse
Predictable	Unipredictable
Planned	Spontaneous
Compulsory	Voluntary
Anonymous	Personal
Vehicle Oriented	People Oriented
Technical Oriented	Relationship Oriented
Government Oriented	Communitiy Oriented
Avoids Conflict	Embraces Confliet
Speed Oriented	Savors the Moment

Fig. 7.3 Characteristics of the traffic world v. the social world (Engwicht, 2005)

essentially local businesses and developments, precious long-distance service capacity is wasted. For the entire highway system to work for both congestion and climate outcomes, highway planners and their community partners must become highly deliberate in both their design and allocation of road space. Public policy needs to support both transportation and land use planners in this change of focus.

Placemaking: Present guidelines for transportation planners unequivocally require that highways be designed to operate at a speed that satisfies nearly all drivers (AASHTO, 2001). Any reduction in speed is considered to be "inefficient," and contrary to the objective to provide operational efficiency, comfort, safety and convenience to the motorist. From the point of view of the traveling motorist, this guidance sounds reasonable, but if VMT reduction is the new goal, elected officials, community leaders and public policy must all combine to must make a new paradigm crystal clear. Policy guidance laying out these VMT goals must then be connected to planning activities and investment decisions.

Speed is the current objective of transportation agencies. Balance requires that solutions be negotiated, not calculated. This requires community engagement and decision-sharing, something transportation agencies are not clear how to do. Despite the lack of policy guidance, traffic engineers are gradually moving away from a "wider, faster, straighter" mentality in highway design to a more customer-based approach that dovetails with smart growth objectives to make compact, VMT-efficient communities possible. They are adopting a new approach to facility planning, called placemaking, that designs for destinations and roads together, not just for traffic.

The principle behind placemaking is that you get what you plan for. If you plan to accommodate more cars and traffic you will get more cars and traffic. Likewise, if you plan facilities to provide people with access to more high quality places, you will get more people and more high quality places. By balancing their interpretation of geometric design guidelines to focus more on helping travelers improve their trip choices, reliability, quality destinations and safety, and less on design speed as the pre-eminent metric of traveler satisfaction, transportation practitioners can support these placemaking efforts. As destinations become designed into facility planning through parcel-based transportation modeling (see below), VMT growth will decline for the reason that more destinations within any given area means fewer miles of travel are needed to get to them.

Traffic diffusion through connected street grids: The traditional practice of collecting traffic and conveying it to increasingly larger and wider roads, usually on the state highway system, is not an efficient use of system capacity. Grid systems that eliminate the need for turning lanes and provide drivers with more route choices, thus reducing backups, provide more capacity than hierarchical road systems. They also promote more efficient travel because they allow travelers to select more direct routes between destinations. State DOTs can actively support smart growth by making it known to developers and local governments that efforts to reduce local infrastructure costs by dumping development-related traffic onto the state highway system are no longer acceptable. Local traffic can best be served through local grids, preserving state highway capacity for its intended purpose of serving regional and statewide travel needs.

Integrated travel demand models: There is new interest by state DOTs in supporting VMT reductions by:

- Improving their own travel models to include alternative land development patterns
- Helping local governments develop simple, integrated local transportation and land use models
- Implementing a statewide, integrated, inter-regional urban model

The integration of transportation and land use in travel models would allow scenario testing of the VMT growth consequences of various land development patterns. This is already happening in Oregon, Utah and California. When such scenario planning is conducted within the context of a public process, the public almost invariably selects a lower VMT-generating development option, although not always the lowest. At the statewide level, a modally-integrated, interregional urban model would allow state transportation districts and statewide offices to better evaluate solutions, including high-speed and conventional passenger rail improvements, and airport expansions, in corridors that would otherwise presumptively be limited to new freeway options.

State DOTs, with their responsibility for statewide transportation systems, their data collection capabilities, and their modeling expertise, can assume a leadership role in the integration of transportation and land use planning. Caltrans, the state transportation agency in California, is financing regional scenario planning initiatives in collaboration with several metropolitan planning organizations (MPOs) and then crossmarketing the results to other areas of the state. The California Transportation Commission (CTC) has issued Regional Transportation Plan Guilelines recommending scenario planning in conjunction with the development of future statewide and regional long-range transportation plans (CTC, 2007). These integrated modeling initiatives hold significant promise for allowing state DOTs to estimate, and manage, future VMT growth.

Investment Strategies

Investing in alternatives to highway travel is an obvious way for transportation agencies to manage VMT growth. However, at the state level, most DOTs do not own and manage transit facilities, and more than 30 states restrict the use of state gas tax revenues to public highways (Katz and Puentes, 2005). All state DOTs own and manage a state highway system, including the federal aid system. The federal aid highway system, including the interstate system, is owned and managed by states. The only exception is roads on federal lands, which are owned and managed by the federal land management agency where the road is located. While county and local governments own, on average, more than half of all public highways, state DOTs usually set the rules for road and street design and also provide local road and street funding assistance (ICE, 2006). Thus, most state DOTs are primarily in the highway management business, and capital investments are keyed to the needs of the highway system.

Flexible Funding Programs: Before 1991, federal transportation assistance was tied to specific types of federal highway programs—such as interstate, primary, secondary, and urban. However, in 1991 Congress enacted the Intermodal Surface Transportation Efficiency Act (ISTEA) which vastly increased the flexibility of federal funding assistance and which directed the states to pay "insistent attention to the concepts of innovation, competition, energy efficiency, productivity, growth and accountability." The federal intent was that states should invest federal transportation funds where they would do the most good to meet national goals. These goals included "improved air quality, energy conservation, international competitiveness, mobility for elderly persons with disabilities, and economically disadvantaged persons in urban and rural areas of the country" (ISTEA, 1991).

While the percentage of road systems in good repair increased after enactment of ISTEA, an analysis of state transportation agency investments for the 10-year period ending in 2002 indicated that states were switching funds from programs to repair bridges to capital road construction and were leaving funding for air quality improvement, public safety and alternative modes of travel, including bicycling and walking, unspent (STPP, 2003). True, VMT was rising rapidly and states were struggling to keep up with the resulting congestion through capacity improvements. However, this approach begs the question whether accommodating more and more traffic is an efficient strategy for dealing with congestion in an era when new national objectives argue for investing in demand-side strategies to remove trips from the system.

Some state DOTs are trying a new approach. New Jersey is using state highway funds to invest in a "Transit Villages Initiative" that makes incentive grants to towns adopting smart growth strategies and concentrating housing and jobs around transit stations. It also provides technical assistance to these towns in brokering assistance from other state agencies to advance their smart growth, VMT-efficient objectives. Massachusetts has developed a Commonwealth Capital Fund that links local funding assistance to local efforts to protect capacity on the state system through smart growth initiatives. Massachusetts also developed a transit-oriented development bond program that provides capital assistance for bicycling and walking improvements close to transit. The Metropolitan Transportation Commission (MTC) in the San Francisco Bay area provides transportation grants to local governments based on the number of housing units located within a quarter mile from a transit station (NGA, 2007).

Investments to Reduce non-work trip VMT: DOTs, for good reasons, tend to focus their demand management efforts on reducing peak hour, work-trip travel, since it is the commute trip that puts the greatest strain on system capacity. However, such efforts often simply redistribute commute trips to the shoulder hours just before and after the peak hours and can even increase total VMT by increasing average travel speeds. This is because the commute trip tends to be inelastic, meaning travelers continue to drive to work regardless of cost and availability of alternatives. The need for reliable travel times, as well as trip chaining (dropping off clothes at the laundry, kids at school etc.) are cited as primary reasons for this inelasticity when travel options are available. Whatever the reason, DOTs tend to seek solutions for accommodating peak hour demand, not moving it off the system.

Recently, researchers have noticed that non-work trips—including shopping, social trips, dining out, and so on—tend to be more elastic, especially when they include high quality destinations and safe travel options, such as connected sidewalks and pedestrian-friendly intersections, within walking or bicycling distance from residences (Kuzmyak, 2006). Traffic engineers, who design and manage about 70 percent of the road network in most states, are now collaborating with public space advocates, the new urbanist community, and bicycle and pedestrian advocates to test the idea that great destinations within pedestrian-friendly environments can move trips off the system. The Institute of Traffic Engineers has published a proposed recommended practice in this area (ITE, 2006). From a carbon dioxide reduction perspective, all VMT is equal, so a greater transportation focus on reducing driving for the more elastic, non-work trip is more likely to achieve more VMT reductions than a sole focus on managing the peak hour commute trip.

Collaboration with Non-Transportation Agencies for Climate and Community Outcomes: Finally, state DOTs can reduce VMT by partnering with nontransportation agencies that promote VMT reduction for reasons independent of congestion relief, such as for air quality, smart growth or climate protection purposes. One easy option is to include CO_2 reduction, and VMT growth targets as policy goals in their statewide transportation plans. California, Massachusetts, New York and Washington State all include climate protection as long-range plan goals (BBG, 2005).

Some MPOs are also moving in this direction. The Metropolitan Transportation Commission, the MPO for the San Francisco Bay Area, has a 10 percent VMT per capita reduction as a goal of its Transportation 2035 plan. The Portland, Oregon Regional Council has had a similar goal in place for 10 years and is making substantial progress towards its achievement. By putting climate protection and VMT reduction in their long range plans, state DOTs are signaling an internal culture change and notifying other agencies that they are ready to have a conversation with other constituencies about what additional primary demands besides traveler convenience are in need of their assistance.

Collaboration with state growth management agencies is an obvious example where cross-agency collaboration to reduce VMT serves common objectives. States with strong grow management laws such as New Jersey, Washington, and Delaware already mandate such collaboration. Two less obvious examples of cross-agency collaboration to reduce VMT include collaboration with state health and human service (HHS) agencies to meet the needs of their constituents for access to health care services, and collaboration with state environmental agencies in encouraging commuters to use alternative modes for the commute trip. By assisting nondrivers in securing access to medical care, and by helping commuters secure the tax benefits of seeking alternative commuting options, state DOTs can develop partnership that result in the removal of vehicle trips from the highway system for mutual public benefit (BBG, 2005).

Pricing Strategies

Road pricing, or tolling, is a relatively new tool in transportation planning circles for reducing VMT. Historically, toll authorities were established to build new highways and thus stimulate VMT. The decision by Congress in 1956 to build the Interstate Highway System using a gas tax financing scheme rather than a toll road approach as recommended by the Eisenhower Administration further stimulated VMT by making access to the road space essentially free.

Recently, road pricing to control congestion, smooth out traffic and remove some trips from the system has gained favor. California and Minnesota have both imposed tolls on existing highways to reduce congestion through conversion of free lanes to high occupancy toll (HOT) lanes. Congestion pricing, where an entire road segment, rather than just one lane, is tolled is another example of road pricing that seeks to align the price of access to the cost of the capacity actually used. The SAFETEA-LU law in 2005 authorized the U.S. DOT to allow pilot pricing of the Interstate System in up to three states, and to assess congestion pricing in up to 15 states. Road pricing is definitely in the future for management and financing of U.S. highways and bridges. However, it is still unclear whether pricing initiatives will be tailored specifically to (1) reduce VMT, (2) increase travel choices through expenditure of toll revenues on alternatives to driving, and (3) promote climate protection and energy conservation or, in the alterantive, will be adopted primarily to finance construction of new road capacity with a resulting stimulating effect on VMT.

Pricing can also be applied to other elements of the transportation system. Seattle, for example, is testing a variable parking fee scheme that prices onstreet parking at whatever level is necessary to keep available parking spaces at 80 percent capacity. This allows drivers to an area to have a reliable expectation of finding a parking space and eliminates cruising for a space that is estimated to represent, at times, more than 75 percent of total neighborhood travel (Shoup, 2005) Pay-as-you-drive (PAYD) car insurance is another way to increase the variable cost of driving and thus reduce VMT while not increasing total travel costs.

Integrated Transportation Solutions (ITS)

Urban traffic congestion presents a unique opportunity for state DOTs to forge alliances with municipal and regional agencies to reduce VMT through a combination of:

- Road/congestion pricing
- Recycling of road user charges into alternative modes
- Coordinated system management such as parking surcharges and transportation demand management (TDM)
- Inter-connectivity between modes, known as Integrated Transportation Solutions (ITS)

In 2003, Transport for London implemented a coordinated ITS program including an eight pound sterling (about $16 U.S. dollars) congestion charge for vehicles entering the city center during business hours, combined with significant improvements in bus service to address congestion, CO_2 and air pollution problems (Evers, 2007). While the strategic goal was to support the national climate protection goal of a 60 percent reduction on CO_2 below 1990 levels by 2050, the more immediate goal was congestion relief. Of equal significance was the policy of requiring each sector to reduce CO_2 in direct proportion to its contribution to such emissions (Fig. 7.4). Since transport in London was responsible for 22 percent of CO_2 emissions, the same percentage of total reductions need to meet the goal was assigned to transport.

Through 2006, the program achieved a 30 percent reduction in congestion, a 20 percent reduction in traffic, and a 16 percent reduction in carbon dioxide emissions within the congestion pricing zone. Total trips were not reduced, but instead shifted to public transport or nonmotorized modes. Bus service

Transport sector's contribution to
CO_2 savings by 2025

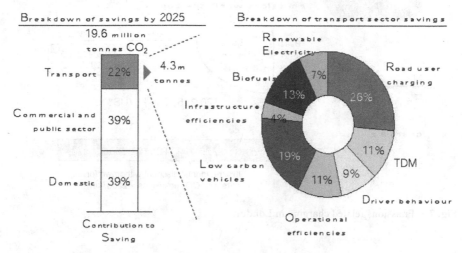

Fig. 7.4 Transport sector's contribution to CO_2 savings by 2025

reliability improved to the point where buses often had to wait at stops because they were ahead of their schedules, car trip times became more reliable, and freight trips became more efficient. Bus patronage went up because riders no longer had to endure long waits at bus stops. Public approval of the congestion pricing system exceeded 70 percent.

The congestion pricing zone was expanded to include western districts of London in 2007. In 2008 a variable, emission-based pricing scheme, where high-CO_2 emitting vehicles will pay higher rates and low-emitters will get credits, will go into effect. Under this scheme vehicles that emit more than 240 grams of CO_2 per kilometer will pay $50 a day to enter the cordon area, while vehicles that emit under 120 grams of CO_2 per kilometer will have free access. Emissions within that range will have varying charges. (See Fig. 7.5). In addition, new commercial buildings will be allowed only one parking space per 12,000 square feet of commercial space. This will add demand management to the ITS initiative.

Transport for London calculates that these measures will reduce transport carbon dioxide emissions by 22 percent. In other words, government is taking lead responsibility for assuring that transportation contributes its proportionate share to CO_2 reductions. According to Mr. Evers, this is a moral calculation, not a calculation of cost-per-ton of CO_2 reduction. Since congestion relief is the primary, near-term objective, any cost is merely a transportation finance charge, with CO_2 reductions a free co-benefit.

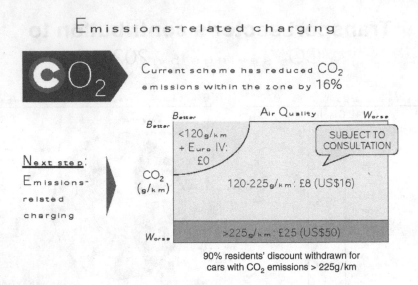

Fig. 7.5 Emissions related charging in London

This integrated solutions approach addresses three transportation challenges facing the US transportation sector simultaneously: (1) urban transportation congestion, (2) finance, and (3) CO_2 emissions. It also finesses two issues of importance in the U.S. carbon reduction debate: cost-per-ton efficiency and the relative responsibility between vehicles, fuels, and system management for addressing transportation carbon. By capping transportation emissions within its jurisdiction through emission charges, funding service improvements through these charges, and implementing demand management initiatives through parking space limitations, London is leading through policy and letting vehicle and fuel carbon efficiency technology fend for itself.

Whether the London congestion pricing program will work in the United States is uncertain. One significant difference between London and most U.S. cities is that, while Transport for London owns and manages all modes of transport under a unified institutional structure, ownership and management of transportation facilities in most U.S. cities is fragmented among local, regional and statewide agencies, transportation districts, toll authorities and other single-purpose transportation institutions. Financing of these facilities is also highly fragmented at all levels of government, including the federal government, with varying rules for securing federal assistance. Absent radical reform of the domestic transportation governance and finance structure, it is not clear whether U.S. transportation practitioners have the ability to adopt ITS solutions similar to London.

Despite these impediments, the U.S. DOT is promoting the London approach through its Urban Partnership Program (UPP), which in August 2007 announced

$850 million in federal grants in support of congestion pricing initiatives. In each area, VMT reductions through congestion pricing are part of the implementation program. For example, New York City intends to reduce city traffic by 6.3 percent under the program. In the San Francisco Bay Area, the MPO has adopted a goal of reducing VMT per capita by 10 percent by 2030. In King County, Washington, which includes Seattle, the county is leading an effort to guide 98 percent of new development to designated urban growth areas, and is committed to building a network of trails and nonmotorized transportation facilities within 0.5 miles of every household. At all levels of government, integrated transportation solutions are being tested. However, it is at the metropolitan level where implementation plans are actually underway.

Conclusion

Transportation agencies are just beginning to understand the confluence of interests between climate protection, congestion relief and smart growth. In a financially and climate constrained world, transportation initiatives that advance broad societal interests as well as providing more reliable, safe travel options will achieve more success in improving total network performance than initiatives that are keyed exclusively to advancing motorist convenience. Integrated solutions, involving partnerships among state agencies as well as regional and local agencies, and which include all three legs of a comprehensive VMT reduction strategy—smart growth, transportation investments and pricing—represent an exciting new approach to transportation problem solving. The key to their success is that they serve multiple customers—including motorists, the community, and declared national policy to achieve mutually beneficial outcomes.

Will these efforts be enough to stave off catastrophic climate change? This question cannot yet be answered. What is clear is that transportation practice must now be cognizant of, and operate within, a climate constrained world. The assumption in this paper has been that the critical target is to keep ambient CO_2 levels below 450 ppm. There is some evidence that this may be too high, and that the critical inflection point is somewhere between 350–400 ppm (we are now at 383 ppm). If this is true, even more dramatic reductions in transportation CO_2 will be required, including absolute reductions in VMT, not simply reductions in VMT per capita. This will require a fundamental rethinking of business supply chains and human settlement patterns to achieve a much higher degree of location efficiency in both areas. In anticipation of this eventuality, the sooner VMT is recognized as a cost of access, not a consumer benefit, the more prepared we will be to address this looming challenge.

Acknowledgement BBG Group LLC. Mr. Burwell would like to acknowledge the assistance of Steve Winkelman of the center for Clean Air Policy, who provided guidance and figures, and Gary Toth of Gary Toth Associates, who provided advice and insight on present transportation planning practice.

References

American Association of State Highway and Transportation Official (AASHTO) 2007a, *Invest in Our Future: A New Vision for the 21st Century*. (Washington, D.C. 2007) at 25. The 1.9% annual VMT growth rate is an FHWA trend rate. In contrast the 2008 Annual Energy Outlook (AOE) published in December 2007 estimates the VMT growth rate at 1. 6% between 2006–2030. These two estimates remain unreconciled.

AASHTO 2007b, Revenue, *Invest in Our Future: Revenue Sources to Fund Transportation Needs* (Washington, D.C 2007) at 13.

AASHTO 2004, *A Guide for Achieving Flexibility in Highway Design*, (Washington, D.C. 2004) at 19.

AASHTO 2001, *A Policy on Geometric Design of Highways and Streets*, (Washington, D.C. 2001) at 66.

BBG Group, *Assessing State Long Range Transportation Planning Initiatives in the Northeast fro Climate and Energy Benefits* 2005 available at http://climate.volpe.dot.gov/papers. html#bbg

Boarnet, M.G. and Crane, R. *Travel by Design: the Influence of Urban Form on Travel*, Oxford University Press, New York, 2001.

Brookings Institution (Brookings 2007) *MetroNation: How U.S. Metropolitan Areas Fuel American Prosperity*, (Metropolitan Policy Program 2007) at 18.

Burkholder, Rex. Presentation at the 2007 Asilomar Conference on Transportation and Climate Change, Asilomar, California, 2007.

California Transportation Commission (2007), "2007 Regional Transportation Plan Guidelines," (adopted September 20, 2007), at 23.

Engwicht, David, "Mental Speedbumps: The Smarter Way to Tame Traffic" Envirobook (Annandale, Australia 2005) at 43.

ENO Transportation Foundation, "Working Together to Address Induced Demand," 2002.

Evers, Mark. presentation at the 2007 Asilomar Transportation Conference. http://www.its. ucdavis.edu/events/outreachevents/asilomar2007/presentations/Day%201%20Session% 203/mark%20evers.pdf

Ewing, R., and Cervero, R., "Travel and the Built Environment." Transportation Research Record, 1780, 87–114, 2001.

Hu, Patricia S. et al., "Projecting Fatalities in Crashes Involving Older Drivers, 2000–2005" Report #ORNL-6963, (Oak Ridge National Laboratory 2000) at 7–1.

Intermodal Surface Transportation Efficiency Act (ISTEA), P.L. 102-240, enacted December 18, 1991. Section 2, Declaration of Policy, codified at 49 U.S.C. 101 note.

Information Center for the Environment (ICE), *Final Report to Caltrans: Assessment of Regional Integrated Transportation/Land Use Models*, Institute for Transportation Studies: Davis, California 2006.

IPCC, *Working Group II Contribution to the Intergovernmental Panel on Climate Change Fourth Assessment report, Summary for Policymakers*, April 2007.

ITE, *Context Sensitive Solutions in Designing Major Urban Throughfares for Walkable Communities*, Institute for Traffic Engineers, (Washington D.C. 2006).

Katz, Bruce and Robert Puentes, ed. *Taking the High Road: A Metropolitan Agenda for Transportation Reform*, Brookings Institution Press (Washington, D.C. 2005) at 72-73.

Kuzmyak, Richard, *Walking Opportunities Index*, presented at the 77th TRB Annual Meeting, January 2006.

Metropolitan Transportation Commission (MTC 2007): "Trends in Vehicle Miles per Capita in Large California Metro Areas 1990-2004" Staff Analysis of USDOT Highway Performance Management System.

National Governors Association (NGA), *State Policy Options for Funding Transportation*, February 2007, available at http://www.nga.org/Files/pdf/0702TRANSPORTATION. PDF

Noland, Robert B. and William A. Cowart, "Analysis of Metropolitan Highway Capacity and the Growth of Vehicle Miles Traveled," paper presented at the 79th Annual Meeting of the Transportation Research Board, Washington, D.C., 2000.

Oak Ridge National Laboratory (ORNL, 2007), Transportation Energy Data Book, 2007, Table 82.

Puget Sound Regional Council, Trends, August 2007.

Shoup, Donald C., The High Cost of Free Parking, (APA Planners Press, 2005).

Surface Transportation Policy Project (STPP), The $300 Billion Question: Are we Buying a Better Transportation System? Washington DC, 2003.

Swope, Christopher, "Rethinking the Urban Speedway," Governing Magazine, October 2005.

Texas Transportation Insititute (TTI), "The 2007 Annual Urban Mobility Report," (College Station, Texas, 2007).

Toth, Gary. "Reducing Growth in Vehicle Miles Traveled: Can we really Pull it off?" in Driving Climate Change: Cutting Carbon from Transportation (Sperling and Cannon editors.) Academic Press, 2007, pp. 129-142.

U.S. Department of Energy/Energy Information Agency (DOE/EIA). Annual Energy Outlook (AOE), Report # DOE/EIA-0383 (December 2007).

U.S. Federal Highway Administration (FHWA), Highway Statistics 2005, 2005.

Urban Land Institute, Growing Cooler; The Evidence on Urban Development and Climate Change, Washington, DC, 2007.

Chapter 8
CO_2 Reduction Through Better Urban Design: Portland's Story

Eliot Rose and Rex Burkholder

Americans are driving more often than ever, and for longer distances. Total vehicle miles traveled (VMT) are growing at 2.5 times the rate of population growth (Ewing, 2007). If current trends continue, the United States (U.S.) will gain 114 million new citizens by the year 2030, with each person driving 16 more miles per day than today (DOT, 2006; Ewing, 2007). According to conventional reasoning, this growth in automobile use is a reflection of consumer choice. Americans simply prefer the independence and personal space provided by automobiles, as well as the access to suburban, large-lot housing that they provide. However, Americans' relationship with automobiles is not a love affair, but a marriage of convenience. Federal policies enacted since World War II have subsidized highway construction, automobile production, and the oil industry, and Americans have reacted sensibly to these policies by buying more cars and driving them more frequently.

Cities have expanded in response to the growth in driving. Land is currently being developed at almost three times the rate of population growth, creating a feedback cycle where drivers must travel farther to traverse sprawling cities (Ewing, 2007). However, infrastructure lasts a long time, and current trends in driving and land use are now butting up against three increasingly harsh realities: climate change, cost, and consumer choice. Growing public awareness of global warming and rising gasoline costs has prompted many Americans to examine their gas consumption more carefully. Hybrid electric car sales are booming in an otherwise sluggish auto market. Between January and July 2007, 49 percent more hybrid electric vehicles were sold in the U.S. than during the same period a year earlier (Associated Press, 2007). Meanwhile, alternative fuels, particularly ethanol, are receiving federal attention.

While lower emissions cars and fuels are certainly an important step in mitigating climate change, a wholesale shift to low emissions, high efficiency vehicles will not be enough to guarantee a sustainable future. Technological improvements will be offset by overall increases in driving, and the environment

E. Rose
Metro, 600 NE Grand Avenue, Portland, OR 97232, USA

D. Sperling, J.S. Cannon (eds.), *Reducing Climate Impacts in the Transportation Sector*, DOI: 10.1007/978-1-4020-6979-6_8,
© Springer Science+Business Media B.V. 2009

will not be able to support the resulting emissions, nor will it be able to support continuing urban consumption of land. Taxpayers will not be able to support the rising costs of infrastructure nor the increase in transportation costs, which increase as cities spread out. Society will not be able to bear the negative effects that car-oriented cities have on health, safety, and social capital.

Reducing transportation's share of greenhouse gas (GHG) emissions will require efforts outside of the transportation sector, particularly in land-use planning. Cities will have to undergo far-reaching changes, and though some of these changes will be difficult to implement, they will deliver benefits that extend far beyond a reduction in emissions. Some of these changes are already underway in the Portland, Oregon, metropolitan region, where per capita GHG emissions have fallen by 12.5 percent since 1990 in the area's most metropolitan county (Portland Office of Sustainable Development, 2005).

The Portland metro region has reduced carbon dioxide (CO_2) emissions while becoming more livable and affordable for its residents. In order to understand the Portland area's successes, it is necessary to understand the relationships between reducing emissions and complementary fiscal and social goals. This chapter examines the policies that have been successful so far in reducing GHG emissions in metropolitan Portland, as well as plans currently being developed to further those successes over the next several decades.

Cutting Emissions and Budgets While Increasing Consumer Choice

Most scientists agree that reductions in GHG emissions between 60 and 80 percent below 1990 levels by 2050 are necessary in order to stabilize climate change, but current trends suggest that GHG emissions, particularly CO_2 from the transportation sector, are only expected to rise. The transportation end-use sector is important because it produces such a significant portion of U.S. GHG emissions. According to the Environmental Protection Agency, the transportation sector is responsible for 33 percent of all CO_2 emissions (EPA, 2007). Accounting for "well-to-wheel" emissions, which take into consideration energy used to produce and distribute fuel as well as fuel use, raises this figure to 43 percent (Replogle, 2007).

Total U.S. non-freight VMT is projected to increase by 1.8 percent annually over the next 10 years, while the average fuel economy of a passenger car is projected to improve by roughly 0.75 percent each year over the same period (EIA, 2007). Therefore, overall gasoline use will continue to rise at a rate of 1.0 percent per year, and the carbon content of fuel is not expected to decrease enough to offset this rise. Even the most stringent feasible standards for fuel economy and low-carbon fuel content, coupled with the most optimistic projections for improvements in automotive technology, will likely be insufficient to even lower greenhouse gas emissions to 1990 levels by 2030 (Ewing, 2007; Greene, 2003, p. 54). A recent study by the Center for Clean Air Policy concludes, "the United States cannot achieve such large reductions in transportation-related CO$_2$ emissions without sharply reducing the growth in miles driven" (Ewing, 2007). One viable strategy to achieve this goal is to arrest the sprawl now occurring at the edges of cities and shorten driving distances between urban destinations.

Policies that aim to reduce GHG emissions often have a difficult time gaining headway because citizens are reluctant to make sacrifices in the present for the sake of future benefits. However, there are also strong financial incentives for smarter urban growth. Both governments and citizens are increasingly unable to bear the costs associated with rising automobile usage. In Oregon, as in most states, the federal government funded 92 percent of all highway construction in the decade following the National Interstate and Defense Highways Act of 1956. The federal share has now dropped to well below 50 percent, but even that level of funding will not last, as many experts fear that the Federal Highway Trust Fund will become insolvent in the next decade. Where federal transportation funding is not enough, local governments are increasingly asking taxpayers to make up the difference.

Even without the burden of extra taxes, transportation costs account for 18 percent of average U.S. household expenditures (U.S. Census Bureau, 2005a). Only housing takes up a larger share of household budgets. As VMT increases and gas prices rise, so will transportation's share of budgets, placing particular strain on low- and middle-income households.

These increasing costs are just one reason that consumers are looking to get away from today's conventional, car-dominated suburbs. Concern over rising obesity rates has increased the demand for housing in pedestrian-friendly neighborhoods. More Americans are responding to the isolation fostered by conventional suburbs by placing increased value on communities that allow for more social interaction. This is particularly true for the growing demographic of homeowners that are single or married without children. Furthermore, people over 65 often prefer not to drive on a daily basis, and as America's largest generation reaches retirement age, there is a rising demand for housing with easy access to goods and services by foot or by transit.

The Portland Area Reins in Greenhouse Gas Emissions

The Portland, Oregon, area has responded to the challenges posed by climate change, cost, and consumer choice by both shortening driving distances between common destinations and providing more efficient modes of transportation. Planners have focused on creating land-use patterns that reinforce transportation goals, so that more people live in areas easily served by transit.

The Portland area is more compact than many metro regions with similar populations because it is surrounded by an urban growth boundary (UGB) that is backed by a strong statewide land-use planning program. The Oregon Land Conservation and Development Commission designates valuable rural natural resource lands and prohibits urban development and services outside of UGBs throughout the state. The UGB is not static. Metro, the Portland area's regional government, is responsible for updating it every five years so that the region's urban area grows along with its population. However, the planning process ensures that expansion happens only if there is a need that cannot be accommodated within the existing UGB, and that good farmland is the last land to be added. Metro requires that newly incorporated land add value to existing regional or town centers, or that the added land becomes a center in its own right.

A 2003 study comparing Portland to four similarly-sized metropolitan statistical areas (MSAs) showed the effectiveness of Portland's UGB in restricting sprawl (Nelson and Sanchez, 2003). Charlotte, North Carolina; Columbus, Ohio; Orlando, Florida; and San Antonio, Texas, all have in between 1.5 and 2.0 million inhabitants in their greater metropolitan areas, while Portland has 2.2 million. San Antonio and Columbus do not have UGBs, and Charlotte and Orlando have UGBs that only apply to the regions' central counties or are not backed by a statewide land-use planning system. Compared to the other four MSAs, the Portland region has a larger urbanized area and more rural land surrounding the city. Between 1990 and 2000, the Portland area added proportionately more densely-populated urban areas, and fewer suburbs and exurbs.

Table 8.1 shows the percentage of overall population growth between 1990 and 2000 that occurred in the urban, suburban, exurban, and rural areas of each MSA. During this time period, 88 percent of Portland's growth occurred in high-density, mixed-use urban areas located close to existing transit lines, jobs and services, compared to 64 percent in San Antonio, 63 percent in Orlando, 31 percent in Columbus, and only 7 percent in Charlotte. As the bottom row of the table shows, Orlando and Charlotte's less stringent UGBs were effective in reducing or restricting growth in rural areas in comparison with boundary-free San Antonio and Columbus. However a greater portion of that growth was channeled into suburbs and exurbs, and less of it into urban areas, when compared to Portland.

UGBs help create more compact, efficient cities that are easier to serve with non-automobile transportation modes. Reliable bus service, streetcar and light rail lines, combined with attention to bicycle and pedestrian planning, ensure that residents who choose not to drive can take advantage of a variety of other travel options. Between 1996 and 2006, per capita annual transit trips in the

Table 8.1 Population growth in Portland and MSAs with similar populations

	Charlotte (%)	Columbus (%)	Orlando (%)	San Antonio (%)	Portland (%)
Urban	7	31	64	63	88
Suburban	50	45	23	8	9
Exurban	45	18	12	12	1
Rural	−1	7	2	17	3

Source: Nelson and Sanchez, 2003.

Portland area grew by almost 20 percent, from 40.8 to 48.9 trips, and transit miles per capita increased by 34 percent, from 156.4 to 210.2 miles (National Transit Database, 2005). Not only is ridership increasing, but residents are also using the system to travel longer distances. There are only six U.S. metropolitan areas with more per capita transit ridership than Portland, and all of them, including New York City and Chicago, have substantially higher populations and a greater portion of their physical layout that dates from before the automobile era (National Transit Database, 2005; APTA, 2007).

Portland's transit network is interlaced with a web of bicycle lanes criss-crossing the city. Many transit and bike facilities also serve pedestrians. This is particularly true in downtown Portland, where four of the six non-freeway bridges over the Willamette River have sidewalks wide enough to accommodate bicyclists and pedestrians side-by-side. Parks and esplanades line both sides of the river, creating a loop that offers easy access to anywhere in the inner central city. Good facilities combined with relatively mild weather make biking easy, and more workers commute by bike in Portland than in any other city— 3.5 percent compared to a national average of 0.4 percent (U.S. Census Bureau, 2005b). Some central neighborhoods boast bicycle commute shares between 5 and 10 percent (Portland Office of Transportation, 2007). Transit and bicycle and pedestrians help get people out of their cars, and also have a positive feedback effect, drawing development in around regional centers and away from the fringes of the city.

One common criticism of smart-growth policies is that they drive up real estate prices, putting too high of a price tag on more sustainable living patterns. However, numerous studies have found no statistical correlation between the Portland area's urban growth boundary and housing prices (Nelson et al., 2002). Oregon's laws require that fast-growing cities like Portland maintain a 20-year supply of land for residential development so that housing supply inside the UGB is not restricted. The Portland area has certainly seen a rapid increase in median home prices, which between 1990 and 2000 grew at twice the rate of median incomes, (Metro, 2003) but the region still has lower median home prices than most other western MSAs with comparable populations (National Association of Realtors, 2007; U.S. Census Bureau, 2006). In a review of academic literature on growth management and housing affordability, Arthur Nelson et al. concluded, "market demand, not land constraints, is the primary determinant of housing prices" (Nelson et al., 2002). If smart growth policies have drawn people to the Portland area and created increased demand for housing, it is a sign that the region is doing something right. The challenge falls to planners and policymakers to ensure that residents of all income levels enjoy the benefits of a livable city, discussed in more detail below.

Rising housing prices in the Portland area have already been partially offset by declining transportation costs. Despite having the same expenditures as the average household in the western states, the average Portland-area household spends 7 percent less on transportation annually, leaving residents more money to spend on housing and entertainment (U.S. Department of Labor, 2005). The average daily commute for a Portland area resident is 20.3 miles, four miles below the national average, and one recent study by economist Joe Cortright estimated that the resulting savings in time, gasoline, and maintenance costs amount to a total of $2.6 billion per year (Cortright, 2007). This money has a value far beyond what the dollar amount would suggest. Since the Portland area does not manufacture cars nor refine petroleum, and residents purchase 10 percent less gasoline than the national average, roughly $800 million that would otherwise leave the region stay in the local economy, stimulating businesses.

Overall, what's good for the Portland area has also been good for the global climate. Bucking national trends, per capita VMT in the Portland area is declining thanks to reliable transit service, smart land-use planning, and outreach programs. Between 1996 and 2000, daily VMT per capita in Portland declined by 6 percent, from 21.3 miles a day to 20 miles a day. So far, the combination of better land-use planning and increased travel options has helped reduce GHG emissions. Metro has yet to conduct a regionwide GHG inventory, but a study in Multnomah County, which is the area's most urban county, showed that per capita GHG emissions have dropped by 12.5 percent since 1990, with almost half of those reductions coming from the transportation sector (Portland Office of Sustainable Development, 2005).

Regional Growth and Reduced Driving Over the Next Three Decades

By the year 2040, the Portland area is projected to add one million new residents, a 47 percent increase over its current population. As the long-term planning agency for the Portland area, Metro is faced with the challenge of continuing to reduce VMT as the region grows rapidly. While current trends certainly are heading in the right direction, much of the gains so far may have come from easily achieved behavioral changes on the part of commuters already living close to centers or transit lines, or younger workers who typically have more flexibility in choosing among different travel options. Continuing to reduce VMT may be difficult, particularly in the suburbs at the fringes of the Portland area.

In 1990, Metro began work on the 2040 Growth Concept, summarized in Fig. 8.1, which identifies regional centers and transportation corridors in which to encourage high-density, mixed-use development in order to guarantee all residents convenient access to employment, retail, and other businesses. Between now and 2035, Metro will invest $1.5 billion toward spurring development in these vital areas, while slowing the expansion of the UGB. This figure may seem large, but it is actually a small share of the overall real estate investments projected to occur in the region over the next three decades. $1.5 billion is just 3.4 percent of $44 billion in projected public investment, and only 0.6 percent of the $260 billion estimate for total investments.

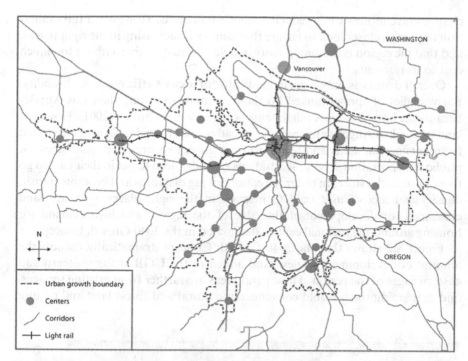

Fig. 8.1 Under the 2040 growth concept, Metro designates centers, corridors, and transit lines along which to focus development over the next several decades
Source: Metro

This is a long-term plan, but it is not a speculative one, thanks to Metro's sophisticated MetroScope modeling software, which allows the evaluation of different investment scenarios and their impacts. Analysts in Metro's Data Resource Center can use MetroScope to compare different planning scenarios across a wide variety of indicators. So far, the predictions that planners have been able to make using MetroScope have proven remarkably accurate. For example, 1996 projections for population growth were within 2.5 percent of today's actual values, and models that predict overall VMT for the Portland area are within 3 percent of the values measured by the state Department of Transportation. One of MetroScope's strengths is its ability to isolate and compare outcomes for a single variable between two scenarios while holding other variables constant. The shading on the following maps produced by MetroScope shows the percentage difference in different variables, for example land consumption and housing demand, between the base-case scenario, with a larger UGB and less investment in regional centers, and the 2040 scenario, with a smaller UGB and higher investment in centers.

Metro's 2040 investments are projected to reduce average travel distances by 5 percent, reduce the average infrastructure needed to build a dwelling unit by 7 percent, and increase the region's overall density by 8 percent. Overall, these

changes save money while reducing vehicle emissions. However, a full evaluation requires a closer look to ensure that density is increasing in the right places,
and that the region is becoming more livable and sustainable without too much
cost to its residents.

Overall density is only a partial indicator of a city's efficiency and livability.
Early studies of sprawl ranked Portland as more sprawling than Los Angeles
because the latter has more inhabitants per square mile (Fulton, 2001). However,
land uses in Los Angeles are generally spread out, with residential areas separated
from commercial areas and few mixed-use areas that are good candidates for
public transportation service, so residents typically need to get in their cars to go
to work or to the store. In order to reduce driving distances and promote transit,
density and mix of uses need to increase in the right places, with less land
consumed for development at the edge of the region, and high demand for
housing around the regional centers designated in the 2040 Growth Concept.

Figure 8.2 shows that the 2040 Growth Concept dramatically reduces the
amount of development on new land added to the UGB at the southern and
eastern edges of the region. These rural areas are farther from existing services,
and developing them would consume agricultural and forest land and require

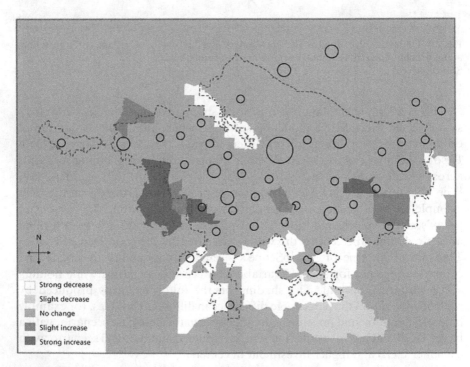

Fig. 8.2 The difference in land developed between the 2040 growth concept and the base-case
scenario
Source: Metro

that residents drive farther to reach their destinations. Meanwhile, newly added land gets developed in selected areas along corridors and near regional centers. The projected growth in land consumption in zones on the western edge of the UGB reflects the growth in the small portion of those lands that are inside the UGB, not growth outside of the UGB.

Figures 8.3 and 8.4 show that demand for both single and multifamily housing shifts to already dense areas inside the UGB, instead of consuming new land. Demand for multifamily housing grows particularly dramatically in the central city. Once again, projected growth in zones straddling the UGB reflects increases in centers within the UGB, not development outside of the UGB. Taken together, these maps show that not only is the total density increasing, but that it's increasing the most in the right places. Metro predicts that under the 2040 Growth Concept, 80 percent of growth will occur within existing urbanized areas in the next 20 years. By investing in centers, Metro is spurring development in places that are close to existing jobs and services, reducing the need for residents to drive and protecting natural resources.

One of the Metro council's goals is for housing to not only be available in mixed-use, walkable neighborhoods, but also be affordable. Under the 2040 Growth Concept, the cost of single-family homes is projected to rise between

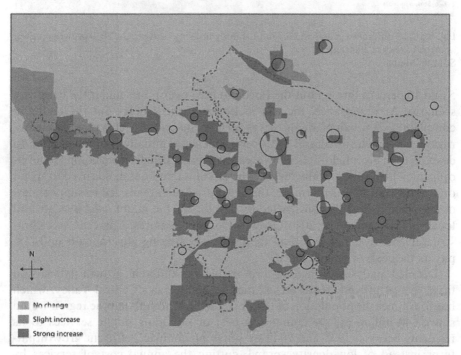

Fig. 8.3 The difference in demand for single-family housing between the 2040 growth concept and the base-case scenario
Source: Metro

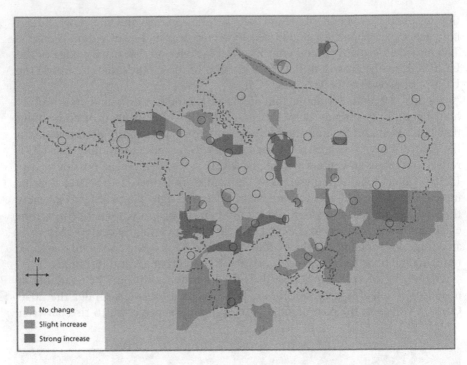

Fig. 8.4 The difference in demand for multi-family housing between the 2040 growth concept and the base-case scenario
Source: Metro

5 and 15 percent throughout the Portland area over prices under the base-case scenario. At first glance, the rise in housing prices bears out the common complaint that urban planning drives up housing prices. Metro plans to encourage development in central locations, which are initially more expensive to develop than greenfield sites. However, all citizens benefit from compact development, which substantially reduces the amount that households need to spend on transportation. Furthermore, the new infrastructure needed to support new development is usually constructed by developers, but maintained by state and local governments. The 2040 Growth Concept requires 7 percent less infrastructure per dwelling unit than the base case, sparing governments and taxpayers the costs of maintenance.

Detailed analysis of the long-term impacts of different growth patterns on taxpayers is difficult. However, a study of California's Central Valley, which has four times the population of the Portland area, found that the region would shave $40 billion off the cumulative cost of providing public services to its residents between 1995 and 2040 by pursuing compact, efficient growth patterns instead of low-density sprawl, cutting the annual cost of services by 19 percent and saving roughly $136 per capita per year in 2006 U.S. dollars (American Farmland Trust, 1995).

In order to fully assess the 2040 Growth Concept's impact on social equity, attention must be paid to the groups most likely to rely on public transportation: low-income, elderly, and single-occupant households. Figure 8.5 shows the difference in housing demand among these demographics between the 2040 Growth Concept and the base-case scenario. With increased investment in the region's centers, demand for low-income, elderly, and single-occupant housing increases substantially in pedestrian-friendly locations with excellent access to transit and retail, particularly in the central city and North Portland. Granting this access is particularly crucial to reducing overall VMT since low-income households that are located in mixed-use, transit-oriented developments in the Portland area are 44 percent less likely to take trips by car than low-income households in the suburbs. In contrast, relocating high-income households in smart growth developments only reduces auto mode share by 17 percent (Metro, 1994). Even though tightening the UGB does lead to an increase in housing prices, it also provides many of the region's less affluent residents with the opportunity to save money on transportation, and saves all taxpayers money that would otherwise be spent on maintaining infrastructure.

By redirecting a small share the overall public investment toward regional centers and tighten the urban growth boundary, Metro should be able to spur

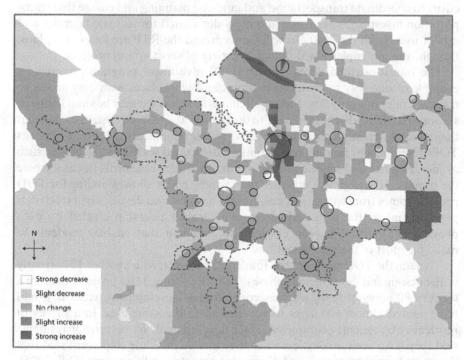

Fig. 8.5 The difference in demand for housing among low-income between the 2040 growth concept and the base-case scenario
Source: Metro

smarter growth, reducing average travel distances in the region by 5 percent and lowering GHG emissions accordingly. Furthermore, increasing numbers of people are drawn to the Portland area precisely because of the "second paycheck" effect of a high quality of life, including increased social equity, reduced infrastructure maintenance costs, and dynamic neighborhoods and urban centers. Although the rise in housing prices appears inevitable, investing in energy efficient development patterns to reduce GHG emissions now seems more prudent than absorbing the high projected costs of adapting to climate change in the future.

Implementing the 2040 Growth Concept

Long-term plans like the 2040 Growth Concept do not stand alone. Other agency projects need to support Metro's 2040 goals, developers need to be able to create smart growth projects without confronting financial barriers, and residents need to understand the transportation options that are available so that they can choose the one that serves them best at the lowest cost. Metro has several programs aimed at realizing the 2040 Growth Concept. In particular, the 2007 update of the regional transportation plan (RTP) represents an across-the-board effort to coordinate transportation and land-use planning and ensure that transportation investments are made in centers designated for increased density and mix of uses. Both the 2040 Growth Concept and the RTP are long-term plans, though, and are implemented though a variety of shorter-term projects.

The transit-oriented development (TOD) investment program provides an example of one way to create new homes and workplaces with easy access to transit. Under this program, Metro purchases land located near bus and light-rail stations and then sells the land back to developers at a reduced cost, provided that they agree to create high-density, mixed-use developments. Metro also assists TOD developers by funding cost premiums associated with higher densities, such as increased fire and seismic protection, and provides easements in cases where the proposed development is denser than zoning codes allow. Funding for TOD projects comes from federal sources, and the amount that development receives is proportional to the projected increases in transit ridership created by each project. The result has been efficient development that enables residents to make the most of their travel options.

To date, the TOD program has funded 21 projects, with another 12 currently in the design and development phases. The 21 existing TOD projects take up a total of 80 acres of land, whereas conventional development patterns would have needed almost 600 acres to accommodate the same uses. In a survey of residents of a recently completed TOD development, 68 percent of residents said they have been driving less since they moved in, while 70 percent said that they now take more transit and 47 percent reported walking more (Dill, 2005).

While the TOD investment program is explicitly geared toward developers, Metro also has programs to help businesses, neighborhood leaders, planners and policy makers create vibrant, mixed-use regional centers. Get Centered! is an outreach program that explains the economic and social benefits of town centers over conventional strip-mall developments and helps interested parties assemble the political, financial and planning tools to create centers. Metro also publishes the Main Streets Handbook, a technical guide for politicians and planners who want to create downtown streets that are attractive to businesses and customers alike. The handbook outlines the pedestrian improvements and zoning codes that support walkable, mixed-use development. Finally, Metro has organized two official visits to Vancouver, British Columbia, Canada, so that Portland area planners, developers, and policy makers can learn what accounted for that city's success in reducing sprawl. These programs help local jurisdictions make investments that enhance community, attract businesses, and draw residents toward destinations that are closer to home and easily served by transit, while the region benefits from reduced congestion and the planet benefits from lower driving-related GHG emissions.

Metro has also established public outreach programs that promote more efficient transportation options. For example, the Bike There! map helps cyclists find the quickest, safest and most pleasant routes around the city, whether they're commuting to work or enjoying a recreational ride. CarpoolMatchNW.org is a free internet service that facilitates carpooling by matching commuters with others in the community who share the same routes. Another website, www.drivelesssavemore.com, provides transit and travel options information, promotes efficient driving practices, and helps users calculate the real cost of driving. Drive Less, Save More staff make regular appearances at public events in order to connect directly with residents. Metro also arranges vanpools for groups of 10–15 commuters, covering 50 percent of monthly costs.

Finally, Transportation Management Associations (TMAs), partially funded by Metro, promote travel options locally in regional centers. One TMA in Northeast Portland decreased single occupancy vehicle trips by 29 percent between 1997 and 2005 by implementing paid on-street parking, improved transit service, and outreach programs promoting biking and walking (Lloyd TMA, 2006). None of these programs call for eliminating automobile use completely, but instead give residents the resources and information needed to make intelligent choices about how to spend transportation dollars.

Challenges from Both Within and the UGB

The progress that the Portland area has made so far in reducing GHG emissions is admirable, but only a small step toward meeting Oregon's statewide goals of bringing total emissions down to 75 percent below 1990 levels by 2050. Even Multnomah County's impressive 12.5 percent per capita CO$_2$ reductions are not enough to offset the county's population growth, and its overall emissions have grown slightly since 2004. Several obstacles need to be to overcome in order to stabilize the changing climate.

Sprawl in Neighboring Cities

As is the case with even the most well-planned cities, the Portland area faces challenges from beyond its urban growth boundary. The ease of long-distance travel afforded by automobiles gives even small towns a footprint that extends far beyond their city limits. As Fig. 8.6 shows, areas beyond Metro's jurisdiction, such as

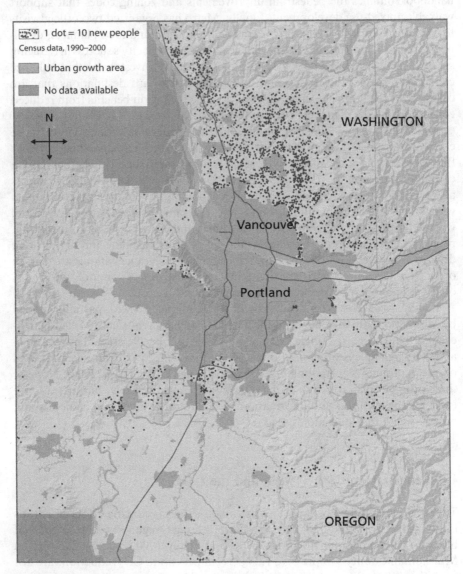

Fig. 8.6 Suburbs, exurbs, and small towns beyond Metro's jurisdiction continue to add residents, mostly in low-density developments
Source: Sightline Institute, 2007 Cascadia Scorecard

Newberg, Oregon, to the southwest, and Clark County, Washington, to the north, continue to spread out onto rural land. This is particularly a concern in mostly suburban Clark County, which has grown at twice the rate of the three Oregon counties in the Portland metro area. Residents of new developments in rural areas are drawn to the Portland area to work and play, commuting long distances to participate in its economy and creating congestion while shirking policies that boost the economy and combat sprawl. In the absence of strong statewide and federal policies to combat sprawl, these development patterns are likely to continue.

Measures 37 and 49

Though many Oregonians appreciate what land-use planning has done for the state, many also bristle at the inflexibility of regulations that protect farm and forest lands. Under intense pressure from groups funded largely by lumber companies that stand to benefit from a relaxation of land-use laws, land-use policy in Oregon has been moving backwards, encouraging sprawl rather than combating it (MIPRAP, 2007; Mortenson and Hogan, 2007). In 2004, Oregon voters passed Measure 37, which entitled property owners to compensation if land use regulations restricted the use of their property and reduced its value. The government could also choose to "remove, modify or not apply" the regulation on a case-by-case basis (State of Oregon, 2004). A total of 7,562 claims were filed under Measure 37, affecting 750,000 acres of land and requesting a total of $20 billion in compensation.

Figure 8.7 shows the claims filed under Measure 37 in areas adjacent to the Portland metro region. The majority of these claims were landowners seeking to subdivide private property into a small number of lots, but a few large landowners, many of them timber companies, sought to create large-scale subdivisions or commercial and industrial developments. In almost every case, governments chose to waive regulations rather than compensate claimants. The majority of Measure 37 claims in the Portland area were far outside the urban growth boundary, creating the potential for longer commute times and new infrastructure on what was once agricultural or forest land.

In November 2007, Oregonians approved Measure 49, an amendment to Measure 37 allowing landowners to build up to three extra homes on their property, but prohibiting commercial development and large subdivisions. Though Measure 49 still facilitates new construction outside of the UGB, Metro estimates that it will produce less than one-sixth the amount of new dwellings that would have been constructed under Measure 37, making it a clearly preferable alternative. Nonetheless, the debate over land-use in Oregon is still far from settled.

Mortgage Policies

Even within cities, current mortgage policies tilt the balance in favor of suburban homebuyers. Loans for homeowners are currently based upon net income, and

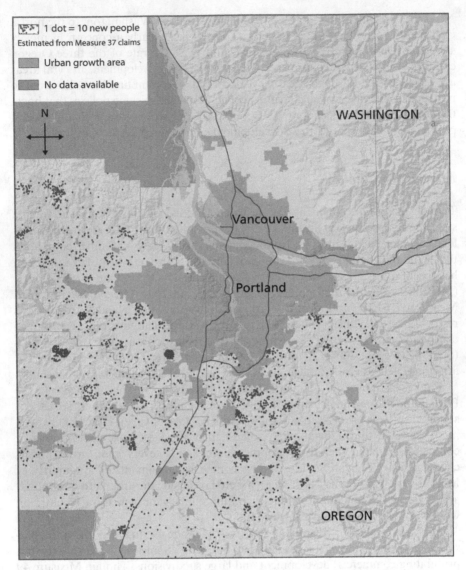

Fig. 8.7 Measure 37, which was overturned by voters in 2007, threatened to create new residential developments on forests and farmlands well outside the UGB
Source: Sightline Institute, 2007 Cascadia Scorecard

they do not take into account cost-of-living expenses. Yet a suburban family typically spends more on transportation than an urban family. The exact size of the difference varies from city to city, but a recent study by the Centers for Transit Oriented Development and Neighborhood Technology showed that transportation costs for suburban households are double those of urban households in the Minneapolis-St. Paul, Minnesota area (CTOD and CNT, 2007).

Home prices tend to be lower in the suburbs, and since the extra money that suburbanites spend on gas and vehicle maintenance does not affect their mortgage rates, they have greater home-buying power than their urban counterparts.

Between 2000 and 2006, mortgage brokers in select U.S. markets offered location-efficient mortgages, which counted the money that residents of walkable neighborhoods with good access to transit saved on transportation toward their incomes, qualifying them for larger home loans. The Federal National Mortgage Association, which guarantees most home loans in the United States, withdrew support for the program in 2006, largely because of the difficulty of compiling data for different markets. Now the Center for Neighborhood Technology is creating a new, easier-to-use affordability index with data for the 50 largest U.S. metropolitan areas, which will hopefully spur new location-efficient mortgage products.

Barriers to Transit Service

As in the rest of the United States, many of the Portland area's residential neighborhoods are laid out in a way that makes it difficult to provide good transit service. Portland's inner city grew up and out along streetcar lines, so serving this area with transit in the modern day is easy; today's bus lines simply follow old streetcar routes. By contrast, the suburbs and small towns of the Portland area grew up in the age of the automobile, and even with good transit coverage across the region it is more difficult to connect these areas to the transit network in a way that serves everyone and all destinations.

For example, Fig. 8.8 shows the transit lines in Washington County, the fastest-growing county in the Portland metro region. Most bus routes are radial, connecting regional centers with the central city. Fewer lines connect regional centers with each other, and even fewer serve major employers, such as Intel and Nike, which have built large campuses in areas where land is more

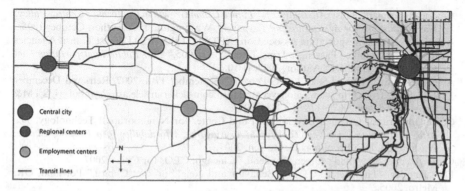

Fig. 8.8 Most transit lines in Washington county connect regional centers with the central city, but few connect centers and residential areas with employment centers
Source: TriMet, 2007

readily available, but transit is not. As a result, many residents who live within five miles of their jobs would still face two or three bus transfers were they to commute by transit, compared to a brief auto commute. Because the infrastructure and zoning in these areas favors automobiles, TriMet, the local transit agency, is reluctant to provide more service because it expects ridership to be low. Better transit service alone cannot solve this problem, nor can better land-use planning. Only a combined effort to create development in centers and serve these centers with increased transit service will work.

Conclusion

Projects like the 2040 Growth Concept provide a picture of the type of comprehensive planning that is necessary to reduce transportation's share of GHG emissions over the long term. It is a picture that is at once optimistic and daunting. It is optimistic because redirecting investment and tightening the UGB will reduce driving distances by 5 percent, and daunting because of the host of challenges that stand in the way and because even a 5 percent reduction in VMT still represents a small portion of the overall change needed to achieve climate stabilization.

However, land-use planning has a positive feedback effect that will facilitate future efforts at climate change mitigation. Metro's future long-term plans will build upon the already-efficient development fostered by the 2040 Growth Concept. As residents come to see firsthand the fiscal, social, and health benefits of smart growth, there will be increased support in the region for even bolder efforts.

References

American Farmland Trust, *Alternatives for Future Urban Growth in California's Central Valley: The Bottom Line for Agriculture and Taxpayers*, Washington, DC: Author, October 1995.

American Public Transportation Association (APTA), "Table 4: 20 Largest Transit Agencies Ranked by Unlinked Passenger Trips, Fiscal Year 2005," *2007 Public Transportation Fact Book*, Washington, DC: Author, 2007, p. 10.

Associated Press, "Hybrid Sales Up 49 Percent," September 17th, 2007. Retrieved December 27th, 2007, from http://news.moneycentral.msn.com/ticker/article.aspx?symbol = US:TM& feed = AP&date = 20070917&id = 7484280

Center for Transit Oriented Development and Center for Neighborhood Technology, *The Affordability Index: A New Tool for Measuring the Affordability of a Housing Choice*, Washington, DC: Brookings Institution, 2007.

Cortright, Joe, "Portland's Green Dividend," Chicago: CEOs for Cities, 2007.

Dill, Jennifer, "The Merrick: Travel Behavior and Neighborhood Choice," Portland, OR: Metro, 2005.

Ewing, Reid, Bartholomew, Keith, Winkelman, Steve, Walters, Jerry, and Chen, Don, *Growing Cooler: The Evidence on Urban Development and Climate Change*, Washington, DC: Urban Land Institute, 2007.

Fulton, William, Pendall, Rolf, Nguyen, Mai, and Harrison, Alicia, *Who Sprawls Most? How Growth Patterns Differ Across the U.S.*, Washington, DC: Brookings Institution, 2001.

Greene, David L., and Schafer, Andreas, *Reducing Greenhouse Gas Emissions from U.S. Transportation*, Washington, DC: Pew Center on Global Climate Change, May 2003.

Lloyd Transportation Management Association (TMA), *2006 Annual Report*, 2006.

Metro, "The Portland Region: How Are We Doing?" Portland, OR: Author, March 2003.

Metro, "Metro Travel Behavior Survey Results," 1994.

Money in Politics Research Action Project (MIPRAP), "Donors Who Gave More than Half the Money to Measure 37 Campaigns File over \$600 Million in Claims; Could Earn Windfall on Investment," April 19th, 2007. Retrieved December 27th, 2007, from http://www.oregonfollowthemoney.org/Press/2007/041807%20Release.htm.

Mortenson, Eric, and Hogan, Dave, "Land-use Fight is Oregonians' Fight," *The Oregonian*, October 1st, 2007.

National Association of Realtors, "Metropolitan Area Housing Prices, Single-Family 3rd Quarter 2007," 2007. Retrieved December 27th, 2007, from http://www.realtor.org/Research.nsf/Pages/MetroPrice.

National Transit Database, *National Transit Profiles*, 2005. Retrieved December 27th, 2007, from http://www.realtor.org/Research.nsf/Pages/MetroPrice.

Nelson, Arthur C., Pendall, Rolf, Dawkins, Casey J., and Knaap, Gerrit J., *The Link Between Growth Management and Housing Affordability*, Washington DC: Brookings Institution Center on Urban and Metropolitan Policy, February 2002.

Nelson, Arthur C., and Sanchez, Thomas W., "Lassoing Urban Sprawl," *Metroscape*, Winter 2003, pp. 13–19.

Portland Office of Sustainable Development, *2005 Progress Report on the City of Portland and Multnomah County Local Action Plan of Global Warming*, Portland, OR: Author, 2005.

Portland Office of Transportation, "Table X: Bicycle Commute Mode Share, 1999–2006," 2007.

Replogle, Michael, "Improving Mobility While Meeting the Climate Change Challenge," Presentation at Metro, November 19th, 2007.

Sightline Institute, *Cascadia Scorecard 2007: Seven Key Trends Shaping the Northwest*, Seattle, WA: Author, 2007.

State of Oregon, Measure 37, 2004. Retrieved December 27th, 2007 from http://www.sos.state.or.us/elections/nov22004/guide/meas/m37_text.html.

TriMet, *TriMet System Map*, Portland, OR: Author, 2007.

U.S. Census Bureau, "Average Annual Expenditures of all Consumer Units by Size and Region, 1995 to 2004," 2005. (2005a) Retrieved November 14th, 2007, from http://www.census.gov/compendia/statab/tables/07s0668.xls.

U.S. Census Bureau, 2005 American Community Survey, 2005. (2005b) Retrieved December 27th, 2007, from http://www.census.gov/acs/www/.

U.S. Census Bureau, Metropolitan Statistical Areas Population Estimates, July 1st, 2006. Retrieved December 27th, 2007 from http://www.census.gov/popest/metro.html.

U.S. Department of Labor Bureau of Statistics, "Selected Western Metropolitan Service Areas: Average Annual Expenditures and Characteristics," *Consumer Expenditure Survey, 2004–05*, 2005. Retrieved December 27th, 2007, from http://www.bls.gov/cex/2005/msas/west.pdf.

U.S. Department of Transportation (DOT), Bureau of Transportation Statistics, "Table 1–33: Vehicle-Miles Traveled (VMT) and VMT per Lane-Mile by Functional Class," 2006. Retrieved December 27th, 2007, from http://www.bts.gov/publications/national_transportation_statistics/html/table_01_33.html

U.S. Energy Information Administration (EIA), "Table 50: Light Duty Vehicle Miles Traveled by Technology Type," *2007 International Energy Outlook*, Washington, DC: U.S. Government Printing Office, 2007.

U.S. Environmental Protection Agency (EPA), *Inventory of U.S. Greenhouse Gas Emissions and Sinks: 1990–2005*, executive summary, Washington, DC: U.S. Government Printing Office, 2007, p. ES-7.

Chapter 9
Transportation-Specific Challenges for Climate Policy

Gustavo Collantes and Kelly Sims Gallagher

Oil security and global climate change are two looming transportation policy challenges. While remarkable advances have been made in reducing emissions of tailpipe pollutants known to cause adverse public health effects, much less progress has been made on reducing overall oil consumption and emissions of greenhouse gases (GHGs). United States (U.S.) highway fuel consumption, almost all petroleum, increased 62 percent between 1973 and 2005, from 110.5 to 179 billion gallons (Davis and Diegel, 2007). Overall, the U.S. transportation sector accounts for 33 percent of the nation's carbon dioxide (CO_2) emissions, with over half of that coming from cars and trucks.

Policy Principles and Criteria

A number of principles or criteria should guide the formation of new federal policies for the transportation sector to address global climate change and U.S. oil dependence (Gallagher et al., 2007). It is important to note that these criteria can be applied to individual policy measures or to packages of measures. Some of the criteria may be highly compatible with each other, while others may be in tension. These include the need for a clear, long-term policy signals versus the need to retain flexibility to change policies in the face of new information. The criteria listed below are approximately listed in order of priority, though all are important. Individually, or in combination, policies should:

- Address both the oil consumption and climate change challenges. Certainly policies should not be adopted that make one of the problems worse, while trying to solve the other. Committing to do no harm could be considered the Hippocratic Oath of energy policy. In addition, policies should make an appreciable difference in addressing one or both problems.

G. Collantes
John F. Kennedy School of Government, Belfer Center for Science and International Affairs, 79 John F. Kennedy Street, Cambridge, MA 02138, USA

D. Sperling, J.S. Cannon (eds.), *Reducing Climate Impacts in the Transportation Sector*, DOI: 10.1007/978-1-4020-6979-6_9, © Springer Science+Business Media B.V. 2009

- Provide a clear, long-term signal to industry and the public. Because industry needs time to alter its production cycles, and because consumers need to make informed purchasing decisions, it is important that the policies provide clear and consistent guidance to the market.
- Be transparent, verifiable, and enforceable. Policies should strive to be transparent to the public and industry in order to better provide the clear long-term signal that they need. In addition, they must be verifiable and enforceable.
- Promote shared responsibility for addressing the problems. The responsibilities for tackling climate change and oil security should be shared among transportation-related industries, including oil companies, auto manufacturers, and biofuels producers. In addition, the burden should be shared by both producers and consumers.
- Protect and assist lower-income segments of U.S. society. Ideally, policies will help lower-income segments of U.S. society, and at worst, they must not harm low-income people.
- Address both fuels and vehicle technologies. Either individually or in combination, policies should induce change in both fuels and vehicle technologies. Approaches that do this are likely to be more equitable and cost-effective than those that load the whole burden onto one side or the other.
- Stimulate innovation. Policies should stimulate technology innovation in order to promote the development of new technologies, and reduce the costs of existing and new technologies so that they enjoy more widespread success in the marketplace.
- Be flexible. Policies should have the capacity to be adjusted in the face of new information and changing circumstances.
- Be cost effective. Efforts should be made to design the most cost-effective policies that are consistent with all of the other criteria presented here.
- Enhance the competitiveness of U.S. industry. To the extent possible, policies should enhance the competitiveness of U.S.-based industry and bolster the U.S. workforce.

There is a large array of policy options for addressing the problems of oil dependence and climate change. Some of these options only offer leverage against one of the two problems, and some offer leverage against both. It is likely that only a portfolio of complementary measures selected from the array, as opposed to any one measure alone, will be able to meet a high proportion of the criteria just outlined.

The Obstacles to Meeting the Policy Goals

There are many policy challenges related to reducing GHG emissions from the transportation sector. Particularly difficult problems in addressing those challenges include various disadvantages of different alternative fuels, the slow turnover of vehicle stock, the slow rate of improvement in new vehicle fuel economy, and conservative consumer responses to new technology and higher fuel prices. Each of these disadvantages is addressed below.

Liabilities of Alternative Fuels

Alternative fuels have important advantages, including energy security, but they also have disadvantages. Even biofuels, which have been received so much recent enthusiasm for biofuels by many analysts and investors, are not a silver-bullet solution to oil dependence or climate change. The fossil alternatives to conventional oil in transport applications—including tar sands, oil shale, and coal-to-liquids technologies—likewise have constraints and liabilities, as does hydrogen, no matter how it is produced.

The new U.S. Energy Independence and Security Act requires a tripling of biofuels production from its 2007 level. Corn prices have almost doubled during the past thirty months, reaching $3.77 per bushel in June 2007, and substantially increasing feed costs for livestock and dairy farmers. Due to the recent surge in demand for ethanol, provoked in part by government subsidies and other mandates, corn farming acreage in the United States is increasing rapidly. According to the National Agricultural Statistics Service, U.S. farmers planted 93.6 million acres of corn in 2007, up from 78.3 million acres in 2006—a 19.5 percent increase. Planted corn acreage in 2007 is the highest since 1944, and state records for planted acreage were set in Illinois, Indiana, Minnesota and North Dakota. Meanwhile, planted area for soybeans fell 15 percent from 2006 levels (USDA, 2007).

Moreover, the use of corn-based ethanol may not result in significant net reductions in either GHGs or energy use. Production can be very energy intensive, depending on factors such as how the corn is grown and then refined into ethanol, how much fossil fuel is used to create chemical inputs like pesticides and fertilizer, and the energy sources used during the refining process. The net energy balance of corn ethanol can be conservatively estimated at about 1.34 (Wang, et al., 2007). Tilman, et al. (2006) determined that the sum of all energy outputs, including co-products, divided by the sum of all fossil energy inputs results in a net energy balance ratio of 1.25 for corn-grain ethanol. Energy and GHG benefits of corn-based ethanol can improve with improvements in refining efficiencies and agricultural practices, particularly increases in crop yields per acre, as shown in Fig. 9.1. Such improvements are not expected to be significant, though.

As a result of the energy intensiveness of its production, involving the use of fossil fuels, corn ethanol averages about 12 percent lower net GHG emissions on a lifecycle basis than gasoline (Wang et al., 2007). Fuel processing strongly affects the GHG content of the fuel on a lifecycle basis. While a new dry mill burning coal produces corn ethanol with no GHG benefits, Wang et al. (2007)

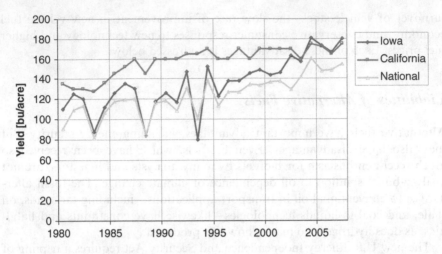

Fig. 9.1 Evolution of corn yield per acre

estimates that even if the biorefinery uses coal instead of natural gas, ethanol is slightly more carbon friendly than gasoline, on a grams of CO_2 per mile basis.

In view of the projected increase in demand in the United States, importation of significant volumes of ethanol will be likely necessary before the full demand can be met with next-generation biofuels. The current law places a cap on the volume of corn ethanol that can used to meet the Renewable Fuel Standard—the rest will need to be met with so-called "advanced biofuel." Advanced biofuel is defined in the Energy Security and Transportation Act of 2007 as "fuel derived from renewable biomass other than corn kernels", which includes ethanol from sugarcane. The volumetric requirements on advanced biofuels under current law and the expected relative economics of the alternatives, will likely make sugarcane—particularly from Brazil—a very attractive ethanol feedstock (Collantes, 2008).

Controlling GHG emissions generated during the well-to-tank stage of the lifecycle process in developing countries constitutes a great challenge. Specifically, a low-carbon fuel standard, as conceived by California and the U.S. Environmental Protection Agency (EPA), seeks to measure the carbon intensity of fuels on a lifecycle basis. Implementing these ideas in a world where ethanol is an international commodity presents many difficulties. Brazil is expected to step up as the main supplier of ethanol to the United States. Preliminary studies in that country suggest that the production of sugarcane and ethanol can be significantly expanded without detriment to the environment (de Carvalho Macedo and Horta Nogueira, 2004). This question is far from resolved though, as many of the potential impacts of sugarcane expansion, including those related to the change in land uses will be indirect and difficult to measure. Some recent evidence suggests that conversion of land for agriculture to supply biofuel production may result in GHG emissions higher than those from the production and use of petroleum fuels (Searchinger, et al., 2008).

Aside from the GHG issue, there are several environmental concerns related to sharply increased production of first generation ethanol, including increased pollution from fertilizers and pesticides, soil erosion from over-reliance on a single crop, and conversion of natural lands into biofuel production. Second generation biofuels, including among others, ethanol produced from cellulosic feedstocks, is still in the research, development, and demonstration stage. While cellulosic biofuels offer the promise of greatly expanded biofuel production from a much wider array of feedstocks, the costs are still very high and the biodiversity and ecological implications remain unclear.

Biodiesel requires much less energy to produce, and so its net GHG reduction as compared to corn ethanol is much better. The net climate impacts of any biofuel depend heavily on production practices, including changes in land use resulting from expanded production. Biodiesel production to date, however, has been limited and expensive. Because biodiesel prices have been higher in Europe, much of the biodiesel produced in the United States has been shipped to Europe.

Production of heavy oil, tar sands, and coal-to-liquid fuels are either already cost competitive or close to competitive, given current crude oil prices, but production of these fuels is very energy and GHG intensive and ecologically destructive (Katzer et al., 2007). Liquid fuels produced from oil shale are not yet economically competitive (Farrell and Brandt, 2006).

Fuel Economy and Vehicle Stock Turnover

The rate of change of the average fuel economy of the on-road vehicle fleet over time is determined by the rate of change of the sales-weighted average fuel economy of each year's new vehicle fleet and the rate of retirement of older vehicles. The fuel economy of new vehicles is regulated in the United States by the Corporate Average Fuel Economy (CAFE) program, adopted in 1975, with enforcement starting in 1978. Congress did not increase the fuel economy standard for passenger cars for 30 years until December 19, 2007, when it passed the Energy Independence and Security Act of 2007. This new law requires an increase in fuel economy for new light-duty vehicles to 35 miles per gallon by 2020.

Average vehicle lifetimes of about 14 years result in slow fleet turnover rates. This effect, together with any fuel economy degradation that may occur over the vehicle lifetime, creates a significant lag in on-road improvement in fuel economy despite the implementation of requirements for increased new-vehicle fuel economy. This lag is shown in Fig. 9.2.

In the United States, the number of relatively high fuel consuming vehicles in the on-road fleet is high in part due to the dramatic rise in sales during the 1990s of sport utility vehicle (SUVs), pick-up trucks, and vans intended for use as passenger vehicles, but subject to the weaker light truck fuel economy standards. These light trucks now comprise 41 percent of registered passenger vehicles in the United States.

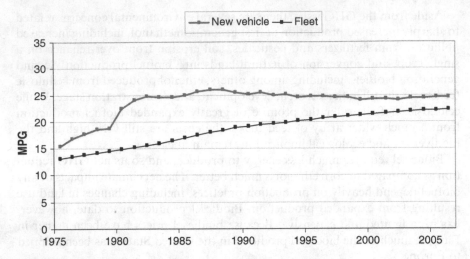

Fig. 9.2 Combined fuel economy for new vehicles and total fleet

The new standards apply to the entire new-vehicle fleets of each manufacturer, thus abandoning the distinction between passenger cars and light-duty trucks in previous CAFE programs. The new CAFE structure induces vehicle manufacturers to consider their vehicle-type mixes as a strategy to meet standards. The modeling discussed in this chapter is specified for passenger cars and light-duty trucks separately. One of the scenarios that we modeled assumed a four-percent per year increase, which closely resembles the new requirements, assuming that new-vehicle fleets are equally divided into passenger cars and light-duty trucks, which is approximately true. Figures 9.3, 9.4, and 9.5 show

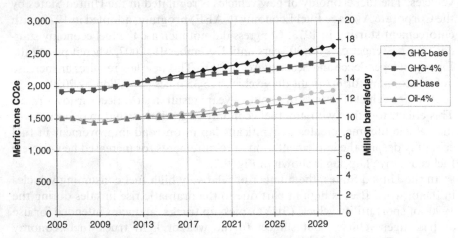

Fig. 9.3 GHG Emissions from transportation and net oil imports resulting from a 4 percent CAFE increase scenario

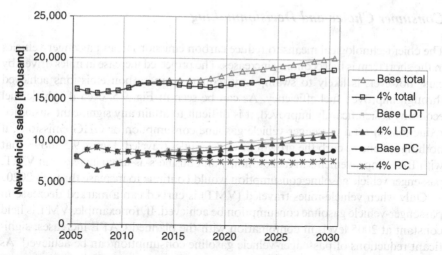

Fig. 9.4 Projected new-vehicle sales, aggregate and by vehicle type, for CAFE increase

the reference business as usual (BAU) base case and the impacts of these scenarios on GHG emissions, oil imports, and new-vehicle sales.

A four-percent annual increase in fuel-economy standards through the year 2020 begins to stabilize GHG emissions from light-duty vehicles. Emissions continue to grow after that point, although initially at a slower rate than in the BAU scenario as vehicle fleet turnover takes place. By the year 2020, emissions in the four-percent scenario amount to 2,185 metric tons of CO_2 equivalent, 109 metric tons less than in the BAU scenario. As the marginal cost of the technologies necessary to meet the standards increase with the fuel economy requirements, so does the average cost of the vehicles, which translates in a slowdown in new-vehicle sales.

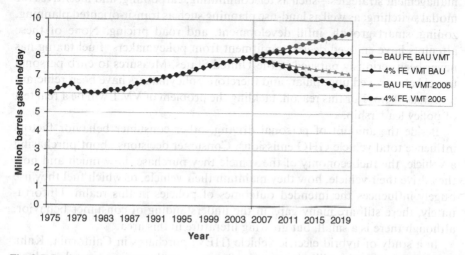

Fig. 9.5 Projections of U.S. passenger vehicle gasoline consumption under different scenarios

Consumer Choice and Decisionmaking

The chief technological means to reduce carbon emissions from passenger vehicles in the short term is fuel economy increases. The projected increase in miles driven by cars, however, is likely to swamp any reductions in carbon emissions achieved through improved fuel efficiency. As can be seen in Fig. 9.5, even if vehicle fuel economy is significantly improved, it is difficult to attain any significant sustained reductions in total passenger-vehicle gasoline consumption or GHG emissions, if nothing is done to curb the growth in vehicle-miles traveled. Figure 9.5 shows that with BAU improvements in fuel economy and business-as-usual increases in VMT, passenger vehicle gasoline consumption would continue to increase through 2020.

Only when vehicle-miles traveled (VMT) is curbed can a marked decrease in passenger-vehicle gasoline consumption be achieved. If, for example, VMT is held constant at 2005 levels in combination with the planned CAFE increases, significant reductions of passenger-vehicle gasoline consumption can be achieved. As shown in Fig. 9.5 fuel-economy improvements alone are not sufficient to meet any meaningful target in carbon emission reductions, for example, reducing emissions to 1990 levels.

Curbing VMT is difficult because of the U.S. population's high dependency on the personal vehicles. Between 1995 and 2005, passenger-car VMT grew on average 1.6 percent each year. SUVs, vans, and light trucks experienced a higher VMT growth rate of 3.0 percent. The *Annual Energy Outlook 2007* reference case prepared by the U.S. Energy Information Administration (EIA) projects a 1.9 percent average annual growth rate for light-duty vehicles under 8,500 pounds through 2030. If the EIA projection is correct, Americans will drive their cars twice as far in 2045 as they drive today.

Attempts to flatten the upward trend of VMT have included travel demand management strategies—such as telecommuting, carpooling, and incentives for modal switching; as well as land-use planning such as transit-oriented planning, zoning, smart growth, infill development, and road pricing. None of these attempts have enjoyed deep commitment from policymakers. Fuel taxing has never been seriously pursued in the United States. Measures to curb personal travel have proved unpopular, and therefore policymakers have been reluctant to pursue them. For this reason, tackling the problem of VMT will be a real test of policy leadership.

Beside the amount of personal driving, other consumer behavior factors influence total vehicle GHG emissions. Consumer decisions about purchasing a vehicle, the fuel economy of the vehicle they purchase, how much and how they drive their vehicle, how they maintain their vehicle, or which fuel they use hugely influences the intended outcomes of policies in this realm. Unfortunately, there still are many gaps in the understanding of consumer behavior, although there is a small, but growing literature in this area.

In a study of hybrid electric vehicle (HEV) purchases in California, Kahn (2007) found that the willingness to pay for more costly environmental products on the part of environmentalists creates market demand for producers that are developing more costly green products. The initial HEV penetration in California occurred predominantly in census tracks with higher percentages of registered green party voters. Later penetration occurred in nearby census tracks that experienced increases in gasoline prices.

Turrentine and Kurani (2007) also found evidence, albeit non-quantitative, through a limited survey of early HEV adopters in California. Many of the interviewees stated that they were primarily motivated by non-economic considerations, such as being a pioneer, an environmentalist, or just "living lighter." In other words, they were not particularly concerned about the specific price difference that they had paid for their HEV. Gallagher and Muehlegger (2007) found in a study of consumer adoption of HEV purchases in the United States from 2000 to 2006 that sales tax incentives, rising gasoline prices and social preferences increased HEV sales 12, 28 and 33 percent, respectively.

A VMT-reduction policy long advocated by many scholars, especially by economists, is fuel pricing or taxing. There is significant uncertainty as to the tax levels needed to affect driving behavior in any meaningful way. Evidence has been published that vehicle travel demand is becoming less sensitive to increases in gas prices. Greene (2000) claimed that structural changes in the demand for gasoline were taking place. This claim was confirmed by Hughes, et al. (2008) who, using monthly data, found that the short-run price elasticity of demand for gasoline in the United States has fallen from -0.21 to –0.34 in the 1975 to 1980 time period to –0.034 to –0.077 in the time period from 2001 to 2006. Using data from the United States in the period from 1997 to 2001, Small and Van Dender (2005) estimated the short-run elasticity at –0.07 and –0.15, depending on the estimation method. These results indicate that consumers today are less responsive to increases in gasoline prices than they were in the 1970s, at least in the short run. It is worth noting that, in the econometric literature, short-run elasticity in fact means instantaneous elasticity: in the case at hand, it means changes in travel during a specific time unit, usually month, quarter, or year, corresponding to a change in gasoline price during the same time period.

Past studies did not include fuel economy among the factors affecting price elasticities of gasoline demand. Conceptually, this omission may be a short-coming because consumers are expected to react to variations in the per-mile cost of driving, which is only partly determined by fuel prices. Fuel economy needs to be included in the specification of a model to estimate the short-run price elasticity of gasoline demand for two reasons. First, the elasticity estimated with a model with specifications that do not include fuel economy will have embedded the effect of any reduction in per-mile driving costs due to increases in fuel economy, and thus it cannot discern the pure-price effect on demand. Second, the price elasticity is not a constant, but rather a function of the per-mile cost of driving. Marginal consumers' reaction to increases in fuel price will vary with the per-mile cost of driving and, therefore, consumers' reaction will be different for different levels of fleet fuel economy. One exception to this modeling shortcoming was done by Small and Van Dender (2005), who studied the effect of changes in per-mile travel cost on the amount of travel, commonly known as the rebound effect. While their focus was on the effect on travel instead of on fuel consumption, they recognized the importance of accounting for the per-mile cost of driving.

In preliminary model runs for the 1978–1982 period, the model showed that including the fuel-economy variable rendered the price elasticity of gasoline demand lower, at about –0.15. In other words, when the per-mile cost of driving is lower, drivers react less to increases in fuel prices. These results lend support to the conceptual expectation that fuel economy is an important variable in analyzing the price elasticity of gasoline demand, and that further work in this area is needed.

For the period 1986–2007, estimates of price elasticity were obtained of about –0.05, while for the period 2000–2007, estimates of about –0.03 were found. These results suggest that the low elasticities that Hughes et al. (2008) assign to recent years may actually be characteristic of the entire period following the oil crises. These estimates did not vary significantly when the evolution in fleet fuel economy was accounted for.

The earlier period, where higher price elasticities were found, was characterized by highly imperfect oil/gasoline markets, with oil shortages and a call by President Carter for Americans to reduce gasoline consumption. The econometric observation of a higher consumer response to gasoline prices in that period may thus be partly an artifact of exogenous factors rather than a pure market response to retail prices.

Longer-run price elasticity is more relevant than the short-run price elasticity for policy because consumer expectations about longer-term future gasoline prices are more likely to cause them to purchase more fuel-efficient vehicles or to make more fundamental lifestyle changes to moderate annual miles traveled. Estimation of the longer-run price elasticity is a more elusive modeling problem, though, which may explain the limited evidence found in the literature. Dahl and Sterner (1991) found that the average long-run price elasticity of gasoline demand was significantly higher than the average short-run elasticity. They found an average short-run price elasticity of –0.26 and a long-run elasticity of –0.86. More recently, Small and Van Dender (2005) estimated the long-run elasticity of fuel consumption with respect to fuel price, for the period of 1997–2001, at –0.34 and –0.64, depending on the estimation method. For the short-run, they obtained estimates of –0.07 and –0.15. The impact of fuel pricing strategies on driving behavior cannot be well understood until more solid and recent evidence on the value of the long-run price elasticity of gasoline demand is developed.

To see the potential impact of transportation-fuel pricing policies, three different policies were modeled, using the Department of Energy's National Energy Modeling System (NEMS):

- A tax of $0.50 per gallon in nominal dollars, starting in 2010
- A tax of $0.50 per gallon on gasoline and diesel, starting in 2010, escalating 5 percent per year in real terms
- A tax of $59 per ton of CO_2 emissions, equivalent to $0.50 per gallon of gasoline, starting in 2010 and constant in real terms over time

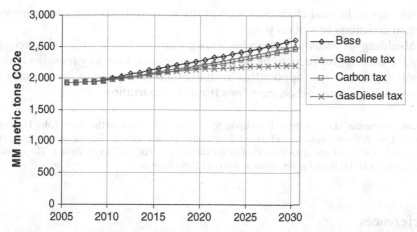

Fig. 9.6 Emissions of GHG from petroleum transportation for scenarios of fuel pricing

Figure 9.6 shows the results for each of these analyses, in terms of GHG emissions. NEMS uses short- and long-run price elasticities of fuel demand of five and 20 percent, respectively. The long-run elasticity used in the modeling is thus lower than the values found in the literature. As expected, a flat $0.50 tax has only a small effect on oil imports and GHG emissions. Levying the tax on the carbon content of the fuel affects 85 percent ethanol mixtures in gasoline (E85) as well, with similar results. This is because the carbon content of E85, per unit of energy, is not much smaller than that of gasoline.

NEMS is not very well equipped to analyze the impact of carbon prices on the market adoption of advanced electric drivetrain platforms such as plug-in HEVs. Therefore, if one is ready to accept that the necessary technologies will be developed within the timeframe considered here, our results for the carbon tax alternative can be considered conservative. Relative stabilization of GHG emissions is likely to be achieved only with a more aggressive taxing scheme, such as the third scenario described above. This variant induces a more meaningful slowdown in VMT increases over time, as well as a stronger adoption of flexible fuel vehicles and consumption of E85. As a consequence, oil imports can be stabilized too.

Final Remarks

The evidence and analysis presented in this chapter are illustrative of the complexity and magnitude of the policy problem facing the United States. The reduction of GHG emissions on the order necessary to avoid significant disruptions in the global climate is challenged by economic, behavioral, and technological factors among others. Climate policies for the transportation sector face additional challenges such as the weak responsiveness of transport carbon emissions to carbon prices.

Modeling of individual policies shows that no single policy is likely to achieve meaningful reductions in carbon emissions. One key message is that a policy package—as opposed to an individual policy tool—is necessary to significantly reduce carbon emissions from transportation.

Acknowledgments This chapter draw upon research conducted together with John Holdren, Henry Lee, Robert Frosch of Harvard University on a recent policy options paper, and Frances Wood and Lessley Goudarzi of OnLocation, Inc. for NEMS modeling. The authors thank Jonathan Hughes for his input in parts of this chapter.

References

Collantes, Gustavo (2008) Biofuels and the Corporate Average Fuel Economy Program: The Statute, Policy Issues, and Alternatives. Discussion paper 2008–05, Cambridge, Mass. Belfer Center for Science and International Affairs. May.

Dahl, Carol and Thomas Sterner (1991) "Analysing Gasoline Demand Elasticities: A Survey." Energy Economics. July.

Davis, Stacy C. and Susan W. Diegel (2007) Transportation Energy Data Book, Edition 26, Oak Ridge National Laboratory, Tennessee.

de Carvalho Macedo, Isaías and Luiz Augusto Horta Nogueira (2004) Avaliação da Expansão da Produção de Etanol no Brasil. Centro de Gestão e Estudos Estratégicos. July.

Farrell, Alexander E. and Adam R. Brandt (2006) "Risks of the Oil Transition," Environmental Research Letters, 1(1), pp. 1–6.

Gallagher, Kelly and Erich Muehlegger (2007) "Giving Green to Get Green? Incentives and Consumer Adoption of Hybrid Vehicle Technology," Discussion Paper, Harvard University.

Gallagher, Kelly, Gustavo Collantes, John Holdren, Henry Lee, and Robert Frosch (2007) "Policy Options for Reducing Oil Consumption and Greenhouse-Gas Emissions from the U.S. Transportation Sector." Discussion paper. Harvard University.

Greene, W. (2000) Econometric Analysis. 4th ed. Upper Saddle River, NJ: Prentice Hall.

Greene, David and Andreas Schafer (2003) Reducing GHG Emissions from U.S. Transportation. Pew Center on Global Climate Change, Washington, DC: May.

Hughes, Jonathan E., Knittel, Christopher R. and Sperling, Daniel (2008) "Evidence of a Shift in the Short-Run Price Elasticity of Gasoline Demand." The Energy Journal, 29(1), January, pp. 113–134.

Kahn, Matthew (2007) "Do Greens Drive Hummers Or Hybrids? Environmental Ideology as a Determinant of Consumer Choice and the Aggregate Ecological Footprint", Working Paper.

Katzer, James, Stephen Ansolabehere, Janos Beer, John Deutch, Denny Ellerman, Julio Friedmann, Howard Herzog, Henry Jacoby, Paul Joskow, Gregory McRae, Richard Lester, Ernest Moniz, and Edward Steinfeld (2007) The Future of Coal: Options for a Carbon-Constrained World. Massachusetts Institute of Technology.

Searchinger, Timothy, Ralph Heimlich, R.A. Houghton, Fengxia Dong, Amani Elobeid, Jacinto Fabiosa, Simla Tokgoz, Dermot Hayes, and Tun-Hsiang Yu (2008) Use of U.S. croplands for biofuels increases greenhouse gases through emissions from land use change. Sciencexpress, February.

Small, Kenneth and Kurt Van Dender (2005) The Effect of Improved Fuel Economy on Vehicle Miles Traveled: Estimating the Rebound Effect Using U.S. State Data, 1966–2001. University of California Energy Institute, Paper EPE-014.

Tilman, David, Jason Hill, and Clarence Lehman (2006) "Carbon-Negative Biofuels from Low-Input High-Diversity Grassland Biomass. Science, 314, p. 1598

Turrentine, Tom and Ken Kurani (2007) "Car Buyers and Fuel Economy," Energy Policy,35, pp. 1213–1223.

USDA (2007) "Acreage," U.S. Department of Agriculture, Washington, DC, June 29, available at http://www.usda.gov/nass/PUBS/TODAYRPT/acrg0607.txt

Wang, Michael, May Wu, and Hong Huo (2007) "Life-Cycle Energy and Greenhouse Gas Emission Impacts of Different Corn Ethanol Plant Types. Environmental Research Letters, 2, April-June, p. 024001

Chapter 10
Are Consumers or Fuel Economy Policies Efficient?

Carolyn Fischer

Recent increases in oil prices, concern about energy security, and apprehension over global climate change have turned attention to fuel economy policy in the United States (U.S.). The primary mechanism to reduce oil use in the U.S. is the set of corporate average fuel economy (CAFE) standards. Paralleling current concerns more than three decades ago, the U.S. Congress was worried in 1975 about increasing imports of crude oil, especially from politically and militarily unstable parts of the world. One response was the Energy Policy and Conservation Act of 1975, in which Congress mandated for the first time that passenger cars and so-called light-duty trucks—pickup trucks, minivans, and sport utility vehicles (SUVs)—had to meet fleetwide CAFE fuel economy standards.

Congress itself set the target for passenger cars at 27.5 miles per gallon (mpg), which translates into 8.6 litre per 100 km of driving, nearly double the pre-1975 average. The National Highway Traffic Safety Administration (NHTSA) was given the responsibility of setting fuel economy targets for light-duty trucks, which was recently increased from 20.7 mpg—a nearly 50 percent increase over 1975—to 22.2 mpg by 2007.

Working in concert with sharply increasing gasoline prices in the early years of the program, the CAFE standards resulted in significant improvements in fuel economy for both passenger cars and light-duty trucks. As a consequence of conservation measures in transportation and other sectors, between 1977 and 1986, imported oil fell from 47 percent to 27 percent of total oil consumption. However, since 1986, fuel consumption rates have been rising again, due to a combination of low gas prices, the plateauing of CAFE standards, and a general shift from cars to SUVs. The recent spike in gasoline prices has prompted a public call for an increase in CAFE standards—and possibly a reform of the program. The Energy Independence and Security Act of 2007 responded by increasing the national fuel economy standard to 35 miles per gallon by 2020 and providing options for certain reforms.

C. Fischer
Resources for the Future, 1616 P St. NW, Washington, DC 20036, USA

D. Sperling, J.S. Cannon (eds.), *Reducing Climate Impacts in the Transportation Sector*, DOI: 10.1007/978-1-4020-6979-6_10, © Springer Science+Business Media B.V. 2009

173

In evaluating the costs and benefits of such policy actions, a key question is whether consumers or fuel economy policies are economically efficient. If policies to address the problems associated with fuel consumption are inefficient, they can be altered to improve them. Moreover, if consumers exhibit inefficient behavior in their choice of fuel economy in vehicles, those inefficiencies have important effects on the efficiency of our fuel economy policies. Either way, the value of the current approach to regulating fuel economy in new vehicles as a cost-effective policy depends on whether or not consumers make inefficient choices.

Why Cafe Standards may not be Efficient

An important feature of the current CAFE program is that it requires each manufacturer to meet separate standards for each of its own car and light truck fleets. Many economists point out that the program would be more cost effective if manufacturers were allowed to trade CAFE compliance credits, much like in Europe, where companies can trade carbon permits or green certificates as part of their strategies to meet climate change or other environmental protection goals.

The point of tradable permits or certificates is to provide automakers with an alternate means of compliance. Some manufacturers might, for example, prefer to specialize in the large-vehicle segment of the passenger car or light-duty truck markets because of a comparative advantage they feel they have in manufacturing or marketing such vehicles. Improvements in those vehicle segments beyond what is required can be transferred to show compliance in other vehicles sectors where progress is harder or more expensive to accomplish. They cannot do so now. On the other hand, if an automaker is able to sell 1 million passenger cars that average 26 mpg, it is allowed to sell another million such vehicles averaging 29.2 mpg in order to hit the 27.5 mpg standard. This has resulted in a situation in which at least some carmakers end up producing and selling for little or no profit, or even at a loss, significant numbers of smaller cars or light-duty trucks to enable them to produce the larger cars or trucks on which they make their money.

If fuel economy credits were fully tradable, an automaker would have another option. If it decided that it could not profitably compete in the small-car or light-duty truck market, it could use any fuel economy credits that it had generated in the other segment of the new vehicle market, or it could purchase credits from another automaker that had exceeded its passenger car or light-truck targets in a previous year. Automakers purchasing credits would be those that find it difficult to manufacture and sell enough smaller vehicles to offset their large-vehicle sales. The automakers choosing to sell credits would be those for which exceeding the standard is less expensive than purchasing credits. Both companies would benefit from the exchange. Furthermore, the lower manufacturing costs from better specialization and more effective allocation of technologies for fuel economy will translate into lower prices for consumers. Meanwhile, the overall fleet of passenger vehicles will meet the same fuel economy goals (Fischer and Portney, 2004). These policy innovations are provided for (though not required) in the recent Energy Independence and Security Act of 2007, which allows the Department of Transportation to establish a program of credit trading across manufacturers.

While economists support making CAFE standards tradable, there is still no clear consensus on whether the benefits of raising those standards actually outweigh the costs. Incorporating technologies to improve fuel economy entails its own costs, either in terms of the price of the vehicle or in tradeoffs with other features that consumers may value more, like engine horsepower or vehicle acceleration. Furthermore, while fuel economy improvements would lower overall fuel consumption, they also lower the cost of driving. This could cause a "rebound effect" where cars are driven further because fuel costs decrease. This would reduce the fuel savings and generate more congestion and accidents, which have costs of their own. Since tailpipe emissions are regulated on a per-mile basis, conventional air pollution increases with miles traveled, so another rebound effect of improving fuel economy could be a deterioration in air quality in some areas. Although studies indicate the rebound effect would be relatively minor, these costs all weigh against the benefits of increasing the CAFE standards.

Many economists note that if the sought-after benefits are reductions in greenhouse gas emissions or oil security costs from gasoline consumption, there already exists a gasoline tax to compensate for them in part. Raising that tax would encourage the purchase of more fuel-efficient vehicles and have direct effects of discouraging fuel consumption by all drivers, reducing vehicle miles traveled, and generating ancillary benefits of less congestion, fewer accidents, and better air quality. Alternatively, since cars account for only 20 percent of U.S. carbon emissions and 45 percent of oil uses, broader policies targeting carbon more directly would be more cost effective for combating these problems.

Ultimately, most of these secondary benefits and costs of CAFE are, well, secondary, as they tend roughly to balance each other out. From an efficiency perspective, the real question turns out to be whether the discounted fuel savings from the regulation outweigh the costs, and that turns out to depend on consumers.

Why Consumers may not be Efficient

Since the oil crises, consumers have not demanded fuel economy beyond what the CAFE regulations require. Indeed, over the last century, passenger vehicles have evolved a great deal, in terms of weight, safety and comfort features, but not in fuel economy. Henry Ford's Model T, introduced in 1908, had a fuel economy between 13 and 21 mpg (Ford, 2008). Today, the popular Ford Explorer has about the same fuel economy (Fuel Economy, 2008).

The puzzle is why fuel economy has not improved in concert with other advances in vehicle technology, especially given the evidence from engineering studies that technologies exist that would more than pay for themselves in fuel savings. Several explanations exist. A simple, but not necessarily satisfactory, one is that those technologies may still be emerging and they will indeed be adopted on their own eventually. Alternatively, consumers may just prefer other features—like horsepower, acceleration, size and weight—that run counter to fuel economy. Certainly, these attributes have been increasing over time, and if this explanation is true, CAFE regulations create "opportunity costs," as consumers are forced to give up features they prize more than fuel economy. A third explanation is that, for whatever reason, consumers may systematically undervalue fuel consumption costs at the time they purchase their vehicles.

This last explanation may seem absurd, particularly to some economists, but many experts point to the problems of limited or poor information to explain consumer undervaluation of fuel economy. Most consumers may know how much they spent on their last tank of gasoline and perhaps even how far they went on that tank, but even the economists among them are unlikely to know everything they need to calculate the value of the next mpg of improved fuel economy. To do that, consumers must know how many miles they will drive each year, how long they will hold the car, what gas prices will be, what is their appropriate rate of interest, and how to apply that in making a present discounted value calculation. Added to these problems are uncertainty about the reliability of the data, and "lemons" problems in the resale market that limit the ability to recapture costs from cars sold before the end of their useful lifetime. Finally, vehicle purchase decisions involve many attributes, and consumers only have so much time to weigh all of them. Since fuel costs are a relatively small part of the overall vehicle cost, they may not be worth calculating in the purchase decision. In a look at individual consumer decision making through in-depth interviews, Turrentine and Kurani (2007) found that households do not track the information needed to calculate their fuel costs and potential savings, and heuristics and lifestyle values may have as much influence on their vehicle purchase decisions where fuel economy is concerned.

Clearly, consumers are willing to pay something for better fuel economy, since they save in operating costs, but the real question is, how much? Unfortunately, little solid empirical evidence exists on the extent to which consumers value or undervalue fuel savings, and what does exist ranges widely. For example, in a simple hedonic study of how different attributes determine vehicle price, Espey and Nair (2005) find that fuel costs seem fully valued. Studies by David Greene and others, based on a rule of thumb used by the automakers, assume that consumers expect a 3-year payback from their investments in fuel economy, which translates into a valuation of about one-third of the fuel savings (e.g., Greene et al., 2005). Other studies fall in between, implying that consumers value three-quarters to two-thirds of fuel costs. Dreyfus and Viscusi (1995) find that consumers seem to have relatively high discount rates, between 11 and 17 percent, while Goldberg (1998), using an impressive structural model of vehicle choice, finds that consumers behave as if they hold the vehicle for seven years, about half the average lifetime. A challenge is that few studies look at consumer valuation directly.

Consumer Inefficiency Matters for Policy Efficiency

Whether the benefits of fuel economy regulation outweigh the costs depends critically on whether consumers "value" fuel economy. If they "rationally" recognize the fuel savings they will achieve, they will be willing to pay for improved fuel economy, convincing manufacturers to offer it, and regulation is unnecessary, theoretically at least. If other vehicle features are more important than fuel economy, then regulation can impose a significant burden on consumers in the form of less desirable cars.

If, on the other hand, consumers are not willing to pay more for fuel-saving technologies, then manufacturers will be unwilling to invest sufficiently in them. In this case, fuel economy regulations can be justified in their own right, even ignoring climate and energy security benefits, as they force manufacturers to incorporate technologies that are worthwhile from society's perspective and that would not be adopted in the absence of regulation.

In a recent study, Fischer et al. (2007) analyzed the effects of a tightening of CAFE standards, incorporating not only the direct costs and benefits of fuel saving technologies, but also the other effects not valued by the market. These include both costs related to fuel consumption—such as carbon emissions, oil dependency, changes in gasoline tax revenues—and also mileage-related costs—such as congestion, accidents, and local air pollution. They consider different scenarios of consumer willingness to pay for fuel savings, and also evaluate the potential savings from making CAFE credits tradable.

Table 10.1 summarizes qualitatively the effects of taking different assumptions about consumer preferences. When consumers are willing to pay fully for fuel savings, the market will provide fuel economy improvements in the range of the policy changes the authors consider, so tightening CAFE has little effect in terms of welfare. When consumers are willing to pay only for one-third of the fuel savings in the purchase price of a car, due to myopia or some market failure, the market will provide some fuel economy improvements, but not as much as would be efficient, so additional tightening of CAFE standards can improve welfare. On the other hand, if we assume that consumers would prefer other amenities over fuel savings, the market efficiently provides improvements in those amenities instead of fuel economy, and more stringent regulation forces manufacturers to provide less desired attributes, leaving consumers worse off.

Table 10.1 Consumer preferences and vehicle market outcomes

Assumption about consumer preferences	Business as usual with emerging technologies	Impact of 4 mpg CAFE increase on consumers
Consumers willing to pay for all fuel savings	Market provides 4 mpg increase	Little or no change in surplus
Consumers willing to pay for 1/3 of fuel savings	Market provides some increase	Raises car prices but consumers gain from fuel savings
Conusmers prefer power and other amenities	Market does not increase MPG	Consumers significantly worse off due to "opportunity costs"

Table 10.2 Welfare change from CAFE increase (cents/gallon saved)

Policy and opportunity cost scenario:	Consumer valuation assumption		
	Far sighted (100%)	High discount (75%)	Short horizon (35%)
4 mpg increase, no opportunity costs	0	0	55
4 mpg increase, opportunity costs	−65	−22	54
4 mpg increase & credit trading, opportunity costs	−57	−21	58
Combined car & truck standards, no opportunity costs	0	0	58
Combined car & truck standards, opportunity costs	−54	−19	53
Combined standards & credit trading, opportunity costs	−44	−10	57

Table 10.2 illustrates the main results in terms of the welfare change expected from each policy scenario combined with an assumption about consumer preferences. Increasing CAFE standards only improves welfare if consumers substantially undervalue fuel savings. If opportunity costs are not important, modest increases in CAFE standards either cause no harm, because they will either not be binding, or they will improve welfare. If opportunity costs are important, tightening the standards can impose substantial costs, unless consumers are significantly myopic. Restrictions on trading across vehicle types reduces efficiency, but these effects are quite small relative to the determinants of whether consumers demand more fuel economy. Fuel and mileage externalities also turn out to be relatively unimportant in justifying CAFE.

In general, economists will argue that problems should be tackled as directly as possible, using mechanisms that signal the costs to society, but allow markets the flexibility to respond in the most cost-effective manner. The costs of greenhouse gas emissions from driving are best signaled by a carbon tax on the fuel. The costs of congestion can be signaled by tolls. Other costs like accidents or conventional air pollution, which accrue with miles traveled, can be addressed by a per-mile charge. For example, the cost of auto insurance can be charged per mile driven. Fuel economy regulation may help curb oil consumption, but from an efficiency standpoint, it is best designed to improve the choices of consumers, assuming they need some help.

Thus, there are tradeoffs in inefficiencies. If consumers value most fuel cost savings, then fuel economy regulation is inefficient and a gas tax or carbon price alone will do the job. On the other hand, if consumers are inefficient and do not fully value fuel economy improvements in vehicle purchases, then regulation can improve efficiency in the economic sense. Of course, direct carbon or oil pricing is still necessary to encourage conservation behavior, and fuel economy regulation serves to complement and improve the effectiveness of those pricing policies. Ultimately, striking the right energy policy balance means knowing how much consumer choices need to be rebalanced.

References

Dreyfus, Mark K. and W. Kip Viscusi (1995). "Rates of Time Preference and Consumer Valuations of Automobile Safety and Fuel Efficiency." *Journal of Law and Economics* 38: 79–98.

Espey, Molly and Santosh Nair. 2005. "Automobile Fuel Economy: What is it Worth?" Contemporary Economic Policy, 23(3): 317–323.

Fischer, Carolyn, Winston Harrington, and Ian Parry (2007) "Do Market Failures Justify Tightening Corporate Average Fuel Economy (CAFE) Standards?" The Energy Journal 28(4): 1–30.

Fischer, Carolyn and Paul Portney (2004) "Tradable CAFE Credits," in *New Approaches on Energy and the Environment: Policy Advice for the President*, Richard D. Morgenstern and Paul R. Portney, Ed. Washington, DC: RFF Press.

Ford (2008). www.ford.com (accessed 01/10/08)

Ford Motor Company. "Model T Facts," http://media.ford.com/article_display.cfm?article_id=858 (accessed 10/26/07).

Fuel Economy (2008). www.fueleconomy.gov/feg/bymanu.htm (accessed 01/10/08)

Goldberg, Pinelopi Koujianou (1998). "The Effects of the Corporate Average Fuel Efficiency Standards in the US." Journal of Industrial Economics 46: 1–33.

Greene, David L., Philip D. Patterson, Margaret Singh, and Jia Li. 2005. "Feebates, Rebate, and Gas-Guzzler Taxes: A Study of Incentives for Increased Fuel Economy," Energy Policy, 33(6): 757–775.

Turrentine, Thomas S. and Kenneth S. Kurani (2007). "Car buyers and fuel economy?" Energy Policy 35: 1213–1223.

Chapter 11
Fuel Economy: The Case for Market Failure

David L. Greene, John German and Mark A. Delucchi

The efficiency of energy using durable goods, from automobiles to home air conditioners, is not only a key determinant of economy-wide energy use but also of greenhouse gas (GHG) emissions, climate change and energy insecurity. Energy analysts have long noted that consumers appear to have high implicit discount rates for future fuel savings when choosing among energy using durable goods (Howarth and Sanstad, 1995). In modeling consumers' choices of appliances, the Energy Information Administration (EIA) has used discount rates of 30 percent for heating systems, 69 percent for choice of refrigerator and up to 111 percent for choice of water heater (U.S. DOE/ EIA, 1996). Several explanations have been offered for this widespread phenomenon, including asymmetric information, bounded rationality and transaction costs.

This chapter argues that uncertainty combined with loss aversion by consumers is sufficient to explain the failure to adopt cost effective energy efficiency improvements in the market for automotive fuel economy, although other market failures appear to be present as well. Understanding how markets for energy efficiency function is crucial to formulating effective energy policies (see Pizer, 2006). Fischer et al., (2004), for example, demonstrated that if consumers fully value the discounted present value of future fuel savings, fuel economy standards are largely redundant and produce small welfare losses. However, if consumers value only the first three years of fuel savings, then fuel economy standards can significantly increase consumer welfare. The nature of any market failure that might be present in the market for energy efficiency would also affect the relative efficacy of energy taxes versus regulatory standards (CBO, 2003). If markets function efficiently, energy taxes would generally be more efficient than regulatory standards in increasing energy efficiency and reducing energy use. If markets are decidedly inefficient, standards would likely be more effective.

D.L. Greene
National Transportation Research Center, ORNL, 2360 Cherakala Boulevard, Knoxville
TN 37932, USA

D. Sperling, J.S. Cannon (eds.), *Reducing Climate Impacts
in the Transportation Sector*, DOI: 10.1007/978-1-4020-6979-6_11,
© Springer Science+Business Media B.V. 2009

The chapter explores the roles of uncertainty and loss-aversion in the market for automotive fuel economy. The focus is on the determination of the technical efficiency of the vehicle rather than consumers' choices among vehicles. Over the past three decades, changes in the mix of vehicles sold has played little if any role in raising the average fuel economy of new light-duty vehicles from 13 miles per gallon (mpg) in 1975 to 21 mpg today (Heavenrich, 2006). Over that same time period, average vehicle weight is up 2 percent, horsepower is up 60 percent, passenger car interior volume increased by 2 percent and the market share of light trucks grew by 31 percentage points. Historically, at least, increasing light-duty vehicle fuel economy in the United States has been a matter of manufacturers' decisions to apply technology to increase the technical efficiency of cars and light trucks. Understanding how efficiently the market determines the technical fuel economy of new vehicles would seem to be critical to formulating effective policies to encourage future fuel economy improvement.

The central issue is whether or not the market for fuel economy is economically efficient. Rubenstein (1998) lists the key assumptions of the rational economic decision model. The decision maker must have a clear picture of the choice problem he or she faces. He should be fully aware of the set of alternatives from which to choose and have the skill necessary to make complicated calculations needed to discover the optimal course of action. Finally, the decision maker should have the unlimited ability to calculate and be indifferent to alternatives and choice sets.

Such requirements are extreme, and it is easy to show that real consumers' decision making does not measure up to them. As Sanstad and Howarth (1994) point out, demonstrating that consumers' decision making falls short of the economically rational ideal does not provide useful guidance for policy formulation. The market failure, or market deficiency, must have important consequences and important implications for choices among policies. Indeed, the term market failure is unfortunate because the usual meaning of the word failure conveys a complete inability to perform a function. Market deficiency or imperfections are perhaps better terms. The uncertainty/loss-aversion deficiency of the market for automotive fuel economy qualifies on both counts.

Energy analysts have examined several forms of market failure to explain the high discount rates consumers appear to apply to future energy savings (e.g., Howarth and Sanstad, 1995; ACEEE, 2007):

- Principal agent conflicts
- Information asymmetry
- Transaction costs
- Bounded rationality
- External costs or benefits

The principal agent problem arises when the agent making the choice of energy using equipment is not the user and therefore does not bear the full consequences of the choice. The agent might find it in his or her interest to buy inexpensive, inefficient energy using equipment, whereas the ultimate customer might have preferred a more efficient version. In the motor vehicle market, consumers are largely unaware of the opportunity to increase vehicle fuel economy at a cost by applying energy efficient technologies. Manufacturers make the technology and design decisions for consumers based on their perception of what consumers will pay for.

Information asymmetry occurs when one party to a market transaction possesses knowledge superior to the other. The suppliers of air conditioners, for example, will have better information about their energy efficiency than the buyers. This enables unscrupulous sellers to deceive consumers, resulting in a reluctance of consumers to trust even scrupulous sellers' high efficiency claims. The adoption of fuel economy labeling has undoubtedly diminished the importance of this problem, yet manufacturers still advertise vehicles based on their highway mileage rather than their combined city/highway fuel economy rating.

The classical formulation of the rational economic choice model takes no account of the transaction costs of optimization. These include the time, effort and expense of collecting and processing information. If these costs outweigh the potential benefit of an optimal choice, rational consumers would decline to optimize. Comparing fuel economy numbers is relatively easy but few consumers have the tools to convert those fuel economy numbers into estimates of present value of fuel savings (Turrentine and Kurani, 2005).

The concept of bounded rationality holds that consumers make rational decisions subject to constraints on their attention, resources and ability to process information, including transaction costs; consumers optimize their decisions but imperfectly (Howarth and Sanstad, 1995). Unfortunately, a priori, it has no implication as to whether consumers' market behavior will undervalue or, as some believe (e.g., Espey, 2005), overvalue fuel economy improvements. Externalities occur whenever a transaction generates costs or benefits to a third party not involved in the transaction. Examples are environmental pollution, traffic congestion, and oil security. If the buyer of a car does not consider the national security consequences of his oil consumption or its impact on global climate change, the car he or she chooses will tend to consume more oil and produce more carbon dioxide than is economically efficient from a societal perspective.

All of these forms of market failure can be seen in the market for fuel economy. Clearly, energy use by automobiles produces externalities, including local air pollution, climate-changing GHG emissions and oil dependence (Parry et al., 2007). Information is also imperfect, as will be shown below, despite the presence of a fuel economy label on every new vehicle. The principal agent issue is also present to some extent. Cognitive and behavioral differences from the economic model of rational consumer behavior are also present. Turrentine and Kurani (2005) found that among 57 California households they surveyed, not one had ever estimated the present value of fuel savings as part of its car-buying decision making. With respect to fuel economy, most consumers' decisions are boundedly rational, at best.

   While each of these market deficiencies is potentially important, uncertainties about the cost and value of fuel economy, combined with loss-averse behavior are sufficient to produce a failure of the fuel economy market to optimize. Moreover, the uncertainty/loss-aversion market failure necessarily results in automotive fuel economy significantly below the economically efficient level.

   A variety of uncertainties make the investment in increased fuel economy a risky bet for consumers. Despite labeling, consumers are not sure what fuel economy will actually be achieved in real world driving. They cannot accurately predict future fuel prices any more than experts can. They are not even certain exactly how much driving they will do or how long their car will last. Consumers' preference for the status quo, combined with fuzzy preferences for future savings guarantee loss-averse behavior. Consumers may be rational and as well informed as possible, yet the market will still decline investments in energy efficiency that have positive expected net present value because of the combined effects of uncertainty and loss aversion.

   Although the focus of this chapter is on automotive fuel economy, it seems clear that the market deficiency identified here should apply more generally to the efficiency of all forms of energy using capital equipment, whether purchased by consumers or loss-averse firms. The combination of uncertainty and loss aversion appears to be a ubiquitous barrier to market acceptance of increased energy efficiency.

What Market Failure?

Irrationality, in the common sense of the word, does not play a major role in the uncertainty/loss-aversion fuel economy market deficiency. Consumers and producers can be quite rational, and still the market will fail to provide an economically efficient level of fuel economy. The problem arises from inherent uncertainties about the value of efficiency improvement and the inherently loss-averse behavior of consumers. Moreover, the locus of the fuel economy market deficiency is not entirely on the consumer's side of the equation. Rational risk aversion on the part of manufacturers, such as avoiding costly investments to produce attributes consumers are not likely to value, is also undoubtedly a factor but is not considered in the analysis below (see, Goldstein, 2007, Chapter 6).

   A key aspect of the fuel economy market deficiency is the fact that the value of fuel economy to the consumer is the difference between the present value of future fuel savings and the cost of achieving those fuel savings via technical changes to a vehicle. This central fact has two very important implications. First, the difference between fuel savings and the increase in vehicle price is certain to be smaller than either, and may be very small relative to the price of the vehicle over a wide range of increased fuel economy. Second, if the consumer is uncertain either about the price of higher fuel economy or its present value, he or she will be relatively more uncertain about the difference between the two. These key elements of the fuel economy market deficiency are not unique to fuel economy and may well be a characteristic of all types of energy efficiency improvements to durable equipment.

The second essential element of the fuel economy market deficiency is consumers' loss aversion. The principle of loss aversion holds that individuals evaluate outcomes not in terms of their impacts on their resulting state of wealth, but rather in terms of changes from a reference state of wealth, and that losses are valued more than equivalent gains (Gal, 2006). The existence of loss aversion has been repeatedly demonstrated in experiments and is held to be the explanation of such phenomena as the "equity premium puzzle," the difference in returns between stocks and a risk-free investment such as treasury bills (Thaler et al., 1997). Gal (2006) characterizes it as ". . .the most robust and important finding of behavioral decision theory."

Initial explanations of the widely observed loss aversion of consumers relied on decreasing marginal utility of wealth. If the utility of an additional dollar of wealth decreases as wealth increases, then consumers will always value the potential loss of a dollar more than the potential gain of a dollar. However, Rabin (2000) has shown that the diminishing marginal utility of wealth is not an adequate explanation for consumers' loss aversion. He proved that the diminishing marginal utility explanation of loss aversion implies that an agent with any degree of risk aversion for small scale gambles would necessarily exhibit an absurdly high degree of risk aversion in larger gambles. For example, an agent who would decline a 50/50 bet of losing \$10 or gaining \$11 as a sole consequence of the diminishing marginal utility of income would also turn down a 50/50 bet of losing \$100 or gaining \$1,000,000. Thus, the diminishing marginal utility of income cannot adequately explain consumers' loss aversion in small wagers.

Gal's (2006) theory of choices among risky alternatives is able to explain loss aversion behavior for relatively small risky bets based on two principles: consumers must have a motive to act and consumers' preferences are often not precise. The first principle implies a preference for the status quo. If the "bet" offered a consumer does not have an expected value superior to his or her current state, the consumer will not accept the bet because there is no motive to do so. The need for a motive to act induces a preference for the status quo.

The second principle implies that the bet offered will have to be measurably better than the status quo because the consumer's fuzzy preferences make him or her indifferent to small inducements. Consider a consumer offered two alternatives, the status quo versus a bet with some degree of risk. If the consumer has precise preferences for these two options, then as the attractiveness of the bet, usually the payoff, increases, there will be a tipping point at which the consumer will switch from preference for the status quo to preference for the risky bet, as shown in Fig. 11.1A. If the consumer's preferences are imprecise, there will be a region in which it is not clear which of two options is preferred. Without a clear preference, Gal (2006) asserts, there will be no motive to choose either and so the consumer will prefer the status quo, as shown in Fig. 11.1B. Gal proposes fuzzy boundary lines for the consumers' fuzzy

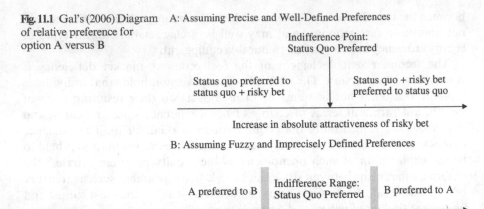

Fig. 11.1 Gal's (2006) Diagram of relative preference for option A versus B

preferences. A more satisfying formulation would be to assume a probability distribution for the consumers' preferences such that the probability of accepting the bet would increase as its attractiveness increases. However, Gal's formulation is sufficient to illustrate the fact that the consumer will require a non-zero premium to accept a risky bet.

Because the consumer has no motive to act unless accepting the bet is an improvement over the status quo, the upper end of the fuzzy preference range must lie to the right of the indifference point in Fig. 11.1A. Thus, as a consequence of imprecise preferences for the status quo versus the risky bet, the consumer will have to be offered a premium over its expected value in order to accept the risky bet. This premium appears to reflect loss aversion, but may instead be explained by the inertia of the status quo and fuzzy preferences. Gal does not claim that his two principles are the only explanation of loss aversion, but rather that they are by themselves sufficient to produce loss aversion.

A rational consumer's preferences for future fuel savings versus an incremental price for energy using equipment must necessarily be fuzzy. Setting aside the uncertainty of the bet itself regarding cost and performance of the equipment, future energy prices, and other factors, unless a consumer knows the future with certainty his or her preferences for future wealth versus present wealth must be fuzzy. Uncertainties about the value of future wealth range from uncertainty about one's future income to uncertainty about whether one will be dead or alive. The corollary is that loss aversion is assured when consumers choose energy using durable goods. Since it is likely that loss aversion for bets that pay off in the future may be different from loss aversion for bets that pay off immediately, understanding how loss aversion varies with the context of a bet could be important to accurately describing loss aversion for future energy savings.

Gal's theory of loss aversion clearly applies to consumers' decisions about fuel economy. The status quo is to not add more expensive technology to a vehicle to increase its fuel economy and thus not receive additional fuel savings.

The status quo is thus known with certainty. The risky bet is to add technology that costs some amount more but may deliver fuel savings of an uncertain amount in the future. If the consumer's preferences are imprecise, the law of inertia states that the consumer will require a premium in expected value of fuel savings to be willing to accept the bet.

In reality, the typical car buyer is not aware of the technological options available for increasing fuel economy because, for the most part, they are not visible in the market at the time of choice. Rather they are choices available to the manufacturer in designing and building future vehicles. The risk would likely increase if the technologies were visible to the consumer due to concerns about the reliability of new technology. If manufacturers believe that consumers are unwilling to accept the risky fuel economy bet, they will decline to adopt the technology or will apply it for other purposes for which they believe the consumer is willing to pay, such as increasing vehicle horsepower or weight.

Consumers' loss aversion has been extensively studied and measured. A well-recognized empirical loss aversion function proposed by Tversky and Kahneman (1992) is the following:

$$u(x) = \begin{cases} x^\alpha & if\ x \geq 0 \\ -\lambda(-x)^\beta & if\ x < 0 \end{cases}$$

where x is the payoff of the bet and u is its utility or perceived value to the consumer. According to Gal (2006), most researchers believe the coefficient of loss aversion, λ, is approximately 2. Typical values for the loss aversion function parameters are $\lambda = 2.25$, with exponents $\alpha = \beta = 0.88$ (Bernatzi and Thaler, 1995). This function will be used in the next section to evaluate the "risky bet" on passenger car fuel economy improvement.

Increasing Passenger Car Fuel Economy from 28 to 35 MPG

The key elements of the economically rational consumer's present value calculation are the fuel economies being compared (E0, E1), the future price of fuel (P_t), anticipated vehicle use over time (V_t), the vehicle's useful life (L), and the consumer's required rate of return (r) for an investment in a depreciating asset, represented by a more efficient car, that produces future revenue in the form of fuel savings. Having estimated the present value, the remaining essential piece of information is the cost of the fuel economy improvement (C). If the uncertainty in these components can be quantified, a probability distribution of net present value can be estimated by Monte Carlo simulation. The net present value (NPV) probability distribution describes the risky bet on higher fuel economy available to the consumer.

Table 11.1 Key parameters of the consumers fuel economy choice problem

Variable	Value Assumed
Miles traveled (first year)	5% = 14,000, mean = 15,600, 95% = 17,200
Rate of decline in usage	4.5%/year
Rate of return required by consumer	12%/year
Vehicle lifetime (extreme value)	mean = 14 years, 5% = 3.6, 95% = 25.3
Gasoline price distribution (lognormal)	5% = \$1.78, mean = \$2.05 , 95% = \$2.63
Incremental price distribution	5% = \$665, mean = \$974, 95% = \$1,385
Fuel Economy Lower	5% = 21 mpg, mean = 28, 95% = 35
Fuel Economy Upper	5% = 28 mpg, mean = 35, 95% = 42
In-Use Fuel Economy Factor	0.85

The assumptions used to quantify each component and its uncertainty are described below and summarized in Table 11.1. Except where stated otherwise, triangular probability distributions were assumed. Two key parameters were assumed to be constants, the consumer's required rate of return was set at 12 percent and the annual decline in vehicle use was set at 4.5 percent. These rates were adopted in the NRC (2002) study of the Corporate Average Fuel Economy (CAFE) standards and are assumed to be known with certainty.

Fuel Economy

There is a fuel economy label on every car, but this label does not necessarily convey to a particular motorist the fuel economy he or she will actually achieve. The sticker even goes so far as to state: "Your actual mileage will vary depending on how you drive and maintain your vehicle." How much will it vary? Almost 15,000 fuel economy estimates voluntarily submitted to the government's website www.fueleconomy.gov indicate a standard deviation of real-world mpg around the adjusted EPA combined city/highway average of 3.7 mpg, as shown in Fig. 11.2 (Hopson, 2007).

Greene et al. (2006) found that the standard error of prediction was reduced by 19 percent if the individual motorist's own estimates of the fraction of driving done under stop-and-go versus freeway conditions were substituted for the EPA's 55/45 percent assumption. On the other hand, correctly computing this number requires calculating a weighted harmonic mean. It is not clear how well consumers can estimate such effects. An individual's knowledge of his or her driving style and experience should also allow a more accurate estimate of in-use fuel economy.

Unfortunately, little if any research exists on how well consumers can predict their own fuel economy based on the EPA estimates. In any case, substantial uncertainty would remain. The two-standard deviation confidence interval of +/−7.4 mpg is rounded to +/−7 mpg. The default assumption is that the two mpg estimates—before and after the increase in mpg—are uncorrelated. An

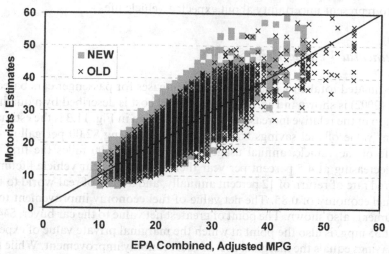

Fig. 11.2 EPA estimated versus motorist estimated fuel economy
Source: www.fueleconomy.gov, Your MPG database, Hopson (2007).

alternative case was analyzed assuming a correlation of 0.5, to represent the possibility that consumers are able to use knowledge of their own driving behavior and traffic conditions to better estimate actual fuel economy.

The future price of gasoline is also uncertain, and quantifying the uncertainty is a challenge. The study assumed that the EIA's high and low oil price cases represent a 95 percent confidence interval for future gasoline prices, and that the reference case represents an expected value. The expected price distribution is lognormal, with mean of $2.05 per gallon, a 5 percent probability of prices exceeding $2.63, a 95 percent probability that prices will exceed $1.78, and a shift parameter of $1.72, indicating a zero percent probability that prices will average below $1.72 per gallon over the vehicle's lifetime. The mean and percentiles correspond to the EIA's reference, high oil price and low oil price cases over the period from 2010 to 2023, discounted according to the consumer's expected rate of return and decline in vehicle use initially at 12 percent and rising 4.5 percent per year.

Even the purchaser of a vehicle does not know precisely how many miles the vehicle will travel in the future. The NRC CAFE committee's (2002) assumptions for the average rate of travel for a new car—15,600 miles per year, declining exponentially at the rate of 4.5 percent per year—were used to calculate expected use. A 95 percent confidence interval of 14,000 and 17,200 was used to describe uncertainty. The NRC CAFE committee assumed an average passenger car lifetime of 14 years. An extreme value survival probability function fitted to scrappage data published in Davis and Diegel (2007, Table 3.8) produced an estimated median lifetime of 14.0 years that was used to represent uncertainty about expected vehicle life.

Incremental Vehicle Price

The estimated total cost of fuel economy increases for passenger cars based on NRC (2002) is shown in Fig. 11.3. Incremental cost is described by a quadratic function of the relative increase in mpg. Also shown in Fig. 11.3 is the estimated present value of fuel savings based on gasoline costing $2.00 per gallon over the life of the vehicle, annual usage beginning at 15,600 miles the first year and decreasing at 4.5 percent per year thereafter, a 14-year vehicle lifetime, a required rate of return of 12 percent annually, and a ratio of real-world to EPA test fuel economy of 0.85. The net value of fuel economy improvement to the consumer is also shown. The point of greatest net value to the car buyer, $444 at about 35 mpg, is also the point at which the marginal private value of expected fuel savings equals the marginal cost of fuel economy improvement. While total costs and benefits climb to $2,500 over the 15 mpg range from 28 to 43 mpg, the net benefits vary between $0 and $444.

Uncertainty about the cost of increased fuel economy is described by the NRC committee's low-cost/high-mpg, average, and high-cost/low-mpg cost curves. The NRC explicitly constructed these curves to approximate a 95 percent confidence interval. The curves were used to predict three estimates of the cost of an increase in passenger car fuel economy from 28 to 35 mpg, and the resulting three estimates were used to define a triangular probability

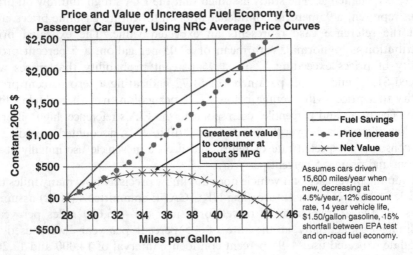

Fig. 11.3 Incremental price, present value of fuel savings and net value of increasing fuel economy to the consumer

distribution with the appropriate expected value and 95 percent confidence interval. The fact that many consumers finance their car payments or lease their vehicles does not change the nature of the consumer's fuel economy choice problem in any fundamental way. The fact that car payments take place in the future does not make the price of increased fuel economy less uncertain.

Consumers never actually see the cost/fuel economy trade-off shown in Fig. 11.3. As a general rule, a car buyer is generally not presented with a menu of fuel economy technologies and their costs to choose from, although some hybrid electric and diesel vehicles are exceptions. For example, a focus group of recent car buyers became confused when they were asked if they would be willing to pay more for a car with better fuel economy. They expected to pay less for cars with higher fuel economy, not more.

A 2002 report by R. Nye at the Looking Glass Group observes that:

> Consumers do not believe they are paying a premium for more fuel-efficient vehicles. They assume there is a direct relationship between the size of the vehicle/engine and fuel economy. They believe that smaller, lighter cars (with the exception of sports cars) are more fuel efficient and are also less expensive. They expect to sacrifice something for better gas mileage (space, comfort, safety, etc.) not pay more for it. (Nye, 2002)

Instead, consumers must compare cars as bundles of attributes and infer the implicit price of fuel economy by trading off the value of several attributes. At any given time, the more fuel efficient version of the same make and model is likely to cost less because it has a smaller engine or a manual transmission. From the consumer's perspective, the cost of fuel economy is the loss of utility associated with a less powerful, less convenient vehicle minus the savings in price. It is the manufacturer who can enumerate the technologies available to increase fuel economy and estimate their costs. Thus, the problem is really that of the manufacturer acting as the car buyer's agent in deciding what technologies to apply to increasing fuel economy. If the manufacturer believes that increasing fuel economy is not a bet the consumer is willing to take, the technologies will not be applied or will be applied for other purposes for which manufacturers believe consumers are willing to pay, such as increasing horsepower or weight.

The present value of future fuel savings is calculated by continuous discounting of the product of the difference in fuel consumption rates, times the price of gasoline, times vehicle miles. The consumer's required rate of return is $r = 0.12$, and the rate of decline in use with age is $\delta = 0.045$.

$$PV = \int_{t=0}^{L} P_t M_o e^{-\delta t} \left(\frac{1}{E_o} - \frac{1}{E_1} \right) e^{-rt} dt$$

The NPV is the PV of fuel savings minus the price increase required to achieve those savings. The problem was programmed in a spreadsheet. Ten thousand simulation runs were executed using the @Risk software. The resulting probability distribution of NPV describes the consumer's fuel economy bet.

Results

The expected net present value of increasing average passenger car fuel economy from 28 to 35 mpg was $405, as shown in Fig. 11.4. However, a substantial proportion of the probability distribution of net value was found to lie on the negative side of zero. There is a small probability that the consumer might lose more than $1,500 on the fuel economy bet. A 90 percent confidence interval ranged all the way from $2,941 down to –$1,556.

Each net present value calculation (x) produced by the simulation is converted to its perceived value, u(x), by inserting it into Tversky and Kahneman's (1992) loss aversion function shown in the equation earlier in this chapter. The loss aversion function described by Equation 1 is graphed in Fig. 11.5. The result is a probability distribution of perceived value, shown in Fig. 11.6.

Figure 11.6 shows the expected value of the bet to the typical, loss-averse consumer to be -$32, implying that the typical consumer would decline to bet on higher fuel economy. Since the consumer is not likely to possess the information necessary to make such calculations, the more correct interpretation is that the manufacturer would decline the bet on the grounds that it would not be attractive to the consumer.

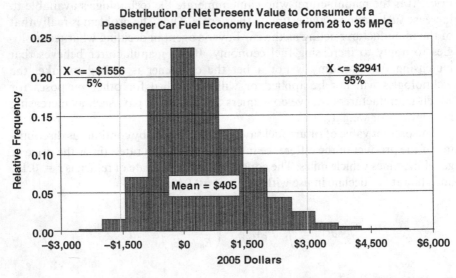

Fig. 11.4 Distribution of net present value to consumer of a passenger car fuel economy increase from 28 to 35 MPG

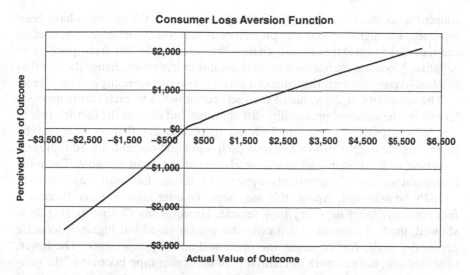

Fig. 11.5 Kahneman and Tversky's loss aversion function

If it is assumed that the lower and upper mpg estimates are correlated with coefficient of correlation 0.5, then the value of the fuel economy bet to the loss-averse consumer becomes slightly positive, $28. While this is still a small number, it does suggest that providing more accurate information to individual consumers about the fuel economy they are likely to realize in the real world can increase the expected value of fuel economy to loss-averse consumers. The key

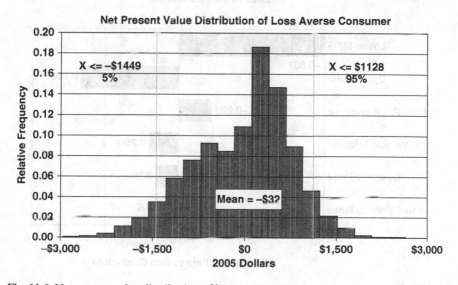

Fig. 11.6 Net present value distribution of loss averse consumer

concept is accuracy, since it appears that the previous EPA estimates have been unbiased, but highly inaccurate, predictors of real world fuel economy. Clearly changes in key assumptions can change the value of the bet from positive to negative. Moderate changes in assumptions, however, cannot change the fact that to a loss-averse consumer the value of a bet on higher fuel economy is close to zero.

The sensitivity of the value of the fuel economy bet to each factor depends largely on the assumed probability distributions, but also on the factor's role in the NPV equation. Figure 11.7 shows regression coefficients obtained by regressing the output variable against each input and normalizing to represent the effect of a one standard deviation change in the input variable. The most important source of uncertainty appears to be the fuel economy that will actually be achieved. Again, this may seem counterintuitive since there is a fuel economy label on every new vehicle. However, as Greene et al. (2006) showed, the fuel economy label estimates are unbiased but highly inaccurate predictors of the fuel economy motorists will actually experience. The lower, or status quo, mpg is more influential than the higher mpg because of the non-linear relationship between mpg and fuel consumption. A 1 mpg change at 28 mpg corresponds to a greater change in fuel consumption than a 1 mpg change at 35 mpg. Net value depends on fuel consumption, which is the inverse of fuel economy.

The next most influential source of uncertainty is the cost of the fuel economy improvement. The importance of vehicle lifetime is frequently overlooked, but if a vehicle is lost due to crashes or severe mechanical failure, the potential for fuel

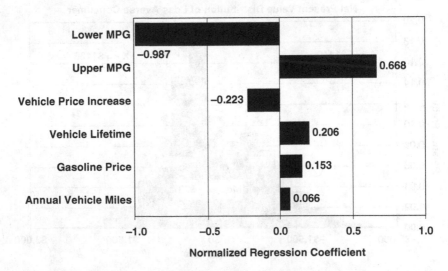

Fig. 11.7 Sensitivity analysis of value of fuel economy increase to loss averse consumer

savings is cut short. Only half of the vehicles will make it to the median expected lifetime of 14 years, which partly explains why vehicle life is an important source of uncertainty.

Unexpectedly, uncertainty about the future price of gasoline is less important than uncertainty about vehicle lifetime. It could also be argued that the EIA's low and high oil price forecasts constitute something less than a 95 percent confidence interval for future fuel prices. Nonetheless, it is clear that even greater uncertainty about future fuel prices would not make fuel price uncertainty the predominant factor in determining the value of the fuel economy bet.

Other Evidence of Fuel Economy Market Failure

The uncertainty/loss aversion market failure has several implications for how markets for fuel economy will function.

- Consumers will appear to undervalue fuel economy.
- Manufacturers will use advances in energy efficient vehicle technology for purposes other than increasing fuel economy, such as increased horsepower, weight or accessories.
- The adoption of fuel economy technology will be relatively insensitive to the price of fuel.
- Governments wanting to significantly increase fuel economy will adopt regulatory standards and/or vehicle fuel economy taxes.

This section provides evidence from fuel economy markets in the United States and around the world that illustrates behavior consistent with the uncertainty/loss aversion market failure.

Consumers Appear to Undervalue Fuel Economy Improvements

When asked what consumers will pay for increased fuel economy, manufacturers generally do not mention uncertainty and loss aversion. They see a potential market failure instead. Manufacturers' perceive consumers' willingness to pay extra for cars in terms of payback periods for fuel economy improvements of 2 to 4 years. Manufacturers' perceptions of consumers' payback periods are based on internal research and studies that are not made public for competitive reasons. However, their perceptions are consistent with what customers say when asked about their willingness to pay for fuel savings in a payback context.

In a survey done for the U.S. Department of Energy (Opinion Research Corporation, 2004), half of the respondents were asked a question about what they would pay for a vehicle that saved $400 per year in fuel, the other half were asked how much they would have to save annually in fuel to justify paying and extra $1,200 for a more fuel efficient vehicle. Payback periods were

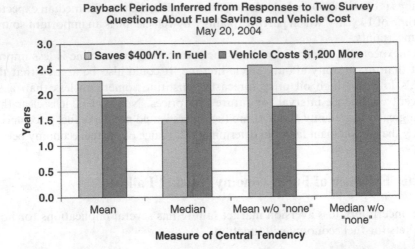

Fig. 11.8 Fuel economy payback periods inferred from a DOE consumer survey

calculated by dividing the mean or median willingness to pay by the $400 in fuel savings, and dividing the mean or median expected fuel savings into the $1,200 additional vehicle cost. The response category "None" was interpreted to mean zero.

The results were remarkably consistent regardless of which question was asked. On average, consumers wanted payback periods between 1.5 and two years, as shown in Fig. 11.8. However, the response category "None" includes both non-responses and consumers who would not pay anything more or would not require any fuel savings. Excluding the response category "None" results in payback periods between 1.8 and 2.6 years. These estimates are considered more correct than those that include the category "None".

As a practical matter, the manufacturers' perception that consumers will pay for 2–4 years of fuel savings may be another way of expressing the consumers' uncertainty and loss aversion. Because of uncertainty and loss aversion, consumers want to recoup their investments quickly. Using the same assumptions used in the analysis shown in Fig. 11.3, but assuming car buyers will pay for only 3 years of fuel savings, produces the graph shown in Fig. 11.9. In light of the uncertainty/loss aversion analysis, the consumers' apparent 3-year payback requirement does not prove that consumers are not rational in the economic sense. The uncertainty/loss aversion explanation implies that they are rational, but recognize the very large uncertainties in the choice they face, and they are loss averse. In either case, the implication is the same: consumers will pay little or nothing for increased fuel economy.

Recent evidence from an in depth survey of the car-buying histories of 57 California households indicates that consumers do not use payback periods to evaluate fuel economy differences when choosing a new vehicle (Turrentine and Kurani, 2005). Nor do they compute the discounted present value of future fuel

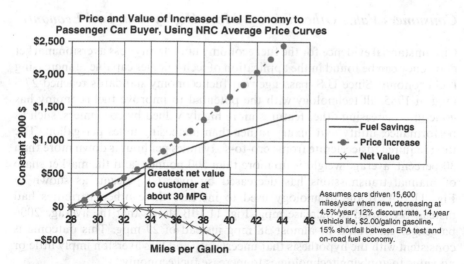

Fig. 11.9 Incremental price, present value of fuel savings and net value of increasing fuel economy to the consumer assuming a simple 3-year payback

savings. Turrentine and Kurani concluded that: "When consumers buy a vehicle, they have neither the tools nor the motivation nor the basic building blocks of knowledge to make a calculated decision about fuel costs."

Indeed, few households mentioned fuel economy when discussing their reasons for the car purchases. When households were asked how much they were willing to pay for a 50 percent increase in fuel economy, only two individuals offered plausible answers based on a payback period. The Turrentine and Kurani study discerned five styles of decision making about fuel economy among their 57 households. Households with limited budgets, like students and enlisted military personnel, deliberately shop for high fuel economy vehicles, but do not consider their annual or long-term fuel costs. Affluent households may have threshold values upon which they base their car purchase decisions with respect to fuel economy. Richer households purchasing luxury vehicles are disdainful of fuel economy and therefore will not consider it. Individuals raised in very poor households are ambivalent about fuel economy, but focus strongly on the price of gasoline. None of the hybrid electric vehicle owners was strongly interested in saving money on gasoline. Their motivations were to protect the environment, to own advanced technology and to be a part of the future.

Turrentine and Kurani's findings were based on a non-representative sample of California households, but they are generally supported by the results of a 1,030 household, 2007 national random sample survey, in which 39 percent of respondents indicated that they did not consider fuel economy at all in their last vehicle purchase (Opinion Research, 2007). Of those who did, only 14 percent mentioned taking economic factors, such as fuel costs or gasoline prices, into consideration.

Consumers Value Other Factors More Highly than Fuel Economy

Circumstantial evidence for the fuel economy uncertainty/loss aversion market deficiency can be found in the application of technologies capable of increasing fuel economy. Since U.S. passenger car fuel economy standards reached 27.5 mpg in 1985, all technology with the potential to improve fuel economy has gone into increasing other features more highly valued by consumers, such as performance, utility and luxury rather than increasing miles per gallon. The time required to accelerate from zero-to-sixty miles per hour is down more than 30 percent, average weight is up more than 500 pounds, and the market share of manual transmissions has decreased by over 60 percent, as shown in Fig. 11.10A. If the technology used to increase these other attributes had instead been used to increase mpg, Fig. 11.10B shows that the average 2006 passenger car would be almost 38 mpg instead of 29 mpg. This outcome is consistent with the hypothesis that uncertainty and loss aversion imply little or no value to applying technologies to increase fuel economy.

Fuel Economy Technology Adoption is Relatively Insensitive to the Price of Fuel

If the uncertainty/loss aversion hypothesis is correct, higher fuel prices will have a muted impact on fuel economy because consumers will appear to not fully value expected fuel savings. This hypothesis can be investigated by comparing fuel prices and levels of technology adoption in the United States and Europe in

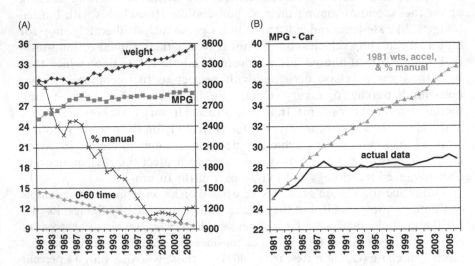

Fig. 11.10 (A) Trends in Passenger Car Weight, Acceleration, MPG and Transmission Type. (B) Impact of Horsepower and Weight on Passenger Car MPG, 1981–2006

the mid-1990s prior to the European voluntary carbon emissions agreement. Based on data from the *International Energy Annual 2005* (U.S. DOE/EIA, 2007), gasoline prices in the United States during this period were just over $1 per gallon compared with $3–$4.50 per gallon in Europe, as shown in Fig. 11.11.

Despite fuel prices roughly three times those in the United States, the application of fuel economy technology in Europe was essentially the same during the 1990s. Table 11.2 compares 4-cylinder engines in the United States and Europe sold in 1993, based on engine descriptions in Ward's World Engines 1993. Although engines produced by U.S. domestic manufacturers were somewhat larger and more powerful, power and torque per unit of engine size were essentially identical. So were compression ratios and the use of overhead camshafts. The U.S. manufacturers led in the percent of engine with 4-valves per cylinder instead of two or three. U.S. manufacturers also led in the use of port fuel injection versus less advanced throttle-body injection or carburetion.

A 2001 report by the International Energy Agency (IEA, 2001) included a detailed comparison of technology use in the United States, Germany, and Denmark in 1998. The market shares of fuel economy related technologies in 1998 compact cars are shown in Fig. 11.12. Three years into the EU's carbon dioxide emissions standards and despite fuel prices roughly three times as high, U.S. technology adoption still matched that of Germany and Denmark. There is evidence, however, that U.S. domestic vehicles lag both European and U.S. imported vehicles in technology adoption. In only three of 18 technologies do the U.S. domestic vehicles have the highest level of fuel economy technology adoption.

A technical appendix to the 2001 IEA report analyzed the differences in fleet average fuel economies across the three countries (IEA, 2000). Fig. 11.13 shows that U.S. passenger cars of comparable size compare well with German and Danish cars. The better average fuel economy of the German and Danish 1998 model year passenger cars is due primarily to differences in vehicle size and performance.

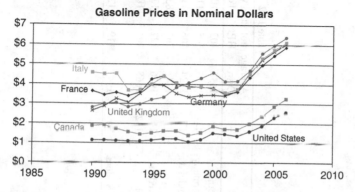

Fig. 11.11 Gasoline prices in North America and Europe, 1985–2006

Table 11.2 Technology indicators for 4-cylinder engines in the U.S. and Europe, 1993

	#	Average (not sales weighted)						% OHC	% 4v	Fuel Metering		
		Disp	HP	Torque	ft-lb/L	hp/L	CR			Port(%)	Single/unk FI(%)	Carb(%)
France	39	1633	102	108	66.0	62.7	9.3	85	13	31	28	41
Germany	44	1755	105	113	64.3	59.6	9.5	100	9	86	5	9
England	17	1724	106	116	67.6	61.7	9.6	88	47	6	71	24
Italy	21	1341	76	83	62.2	56.8	9.2	76	10	33	33	33
Sweden	11	1976	122	134	67.8	61.9	9.4	100	9	45	55	0
All	132	1668	101	109	65.3	60.6	9.4	90	15	48	29	23
U.S.-dom.	34	1947	119	127	65.0	61.0	9.2	91	56	82	18	0

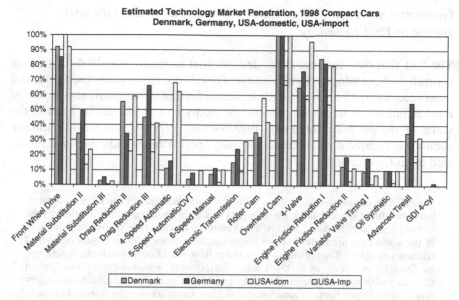

Fig. 11.12 Estimated technology market penetration, 1998 compact cars in the U.S. and two European countries with much higher fuel prices

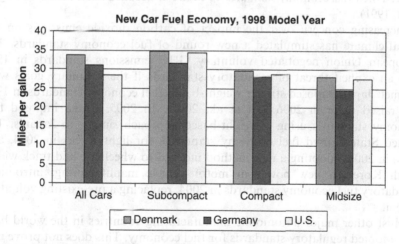

Fig. 11.13 1998 New car fuel economy in Denmark, Germany and the United States

The fact that persistently higher fuel prices in Europe did not lead to noticeably greater adoption of fuel economy technologies in gasoline vehicles is consistent with the uncertainty/loss aversion model of consumers' fuel economy decision making.

Governments Turn to Regulatory Standards to Significantly Increase Fuel Economy

The fact that over the past three decades fuel economy standards in various forms have been adopted by almost every developed economy in the world, and most recently by China and South Korea, is further circumstantial evidence of market failure in the market for fuel economy (An et al., 2007). Even in the wake of the oil price shocks of 1973–74 and 1979–80, the world's leading economies decided it was necessary to adopt regulatory standards. As summarized in an IEA report:

> Nine IEA countries, essentially those with a car manufacturing or assembling industry, have introduced specific standards or targets which are designed to improve directly the fuel efficiency of new passenger cars. These countries account for almost three quarters of the world's car production, which illustrates their overwhelming impact on overall efficiency progress. The policies in place range from mandatory standards, sanctioned by financial penalties, as in the United States, over semi-mandatory ones, as in Canada – with voluntary standards, linked with contingency legislation, to voluntary targets agreed upon between governments and car manufacturers, as in some European countries. (IEA, 1984, p. 16)

However, when world oil prices collapsed in 1986, all of these countries either allowed their fuel economy standards to expire or declined to set increased targets (IEA, 1991).

Increasing concern over the impact of carbon dioxide emissions on the global climate has stimulated a new round of fuel economy standards. The European Union negotiated voluntary GHG emissions standards in 1997 with an explicit threat of mandatory standards if the voluntary goals were not met. Japan imposed stricter weight-based fuel economy standards in 1999 and raised them in 2006 (An et al., 2007). In 2005, China imposed fuel economy standards using a weight-based approach similar to Japan's. The United States raised fuel economy standards for light trucks in 2004, and again in 2006, adopting a new method indexed to wheelbase and track width. South Korea, a new power in motor vehicle manufacturing, introduced mandatory fuel economy standards in 2004, replacing a pre-existing voluntary system.

Most other major automobile manufacturing countries in the world have also adopted regulatory standards for fuel economy. This does not prove that the market for fuel economy is inefficient. However, the adoption of regulatory standards, even by countries with high fuel prices, at least indicates that governments believe that regulatory policies are needed even when gasoline costs $4–$6 per gallon. The evidence presented above is not proof that the uncertainty/loss aversion market failure exists in the market for fuel economy, but it does offer circumstantial evidence of outcomes consistent with such a market failure.

Conclusions

This chapter has proposed a theoretical basis for a market failure, or market deficiency, in the market for fuel economy based on consumers' uncertainty about the value of increased fuel economy and loss aversion. Consumers are known to be loss averse as a general rule. Furthermore, consumers' preferences for future fuel savings versus present wealth are certain to be "fuzzy" as required by Gal's (2006) theory of loss aversion. The net present value of an investment in fuel economy is indeed uncertain, and the uncertainty is enhanced by the fact that net value is the difference of two uncertain quantities.

Manufacturers play a key role acting as the consumers' agents in deciding which technologies should be applied to increase fuel economy. A key premise is that if manufacturers do not believe consumers would consider a bet on increased energy efficiency to be worthwhile, they will not apply advanced technologies to increasing energy efficiency. The risk to manufacturers of betting on increased fuel economy should also be explored.

An example drawn primarily from the National Academy of Sciences' 2002 report on fuel economy standards was used in the study reported in this chapter to quantify the uncertainty in the net present value of increasing passenger car fuel economy from 28 to 35 mpg. The induced probability distribution of net present values constitutes the "bet" available to the consumer. Applying a standard consumer loss aversion function to the net present value distribution produced an estimate of the value of this risky bet to a typically loss averse consumer. The expected value of the increase in fuel economy from 28 to 35 mpg was $405, but when the typical loss aversion of consumers was taken into account, the expected value of the bet on higher fuel economy became –$32. Thus, in the case of fuel economy, uncertainty and loss aversion alone appear to be sufficient to induce an underinvestment, which translates into an under application of available technology to increase fuel economy.

The uncertainty/loss aversion model of consumers' fuel economy decision making implies that consumers will undervalue expected future fuel savings to roughly the same degree as manufacturers' perceptions that consumers demand short payback periods. This suggests that increasing fuel prices may not be the most efficient policy for increasing the application of technologies to increase passenger car and light truck fuel economy. This view is supported by the similar levels of technology applied to U.S. and European passenger cars in the 1990s, despite fuel prices roughly three times higher in Europe. It is also circumstantially supported by the adoption by governments around the world of regulatory standards for light-duty vehicle fuel economy and carbon dioxide emissions.

The implications of the uncertainty/loss-aversion model would seem to extend well beyond the market for fuel economy. Investments in most energy using consumer durable goods share similar characteristics. In addition, to the extent that firms are risk averse, they too might undervalue energy efficient technology. Furthermore, if markets undervalue energy efficient technology, it follows that companies will also undervalue investments in research and development to create new energy efficient technologies. These issues seem to be worthy of further investigation because of their far reaching implications for government policy.

There is evidence that other types of market failures are also present in the market for fuel economy. Nonetheless, it appears that uncertainty plus loss aversion alone are sufficient to induce a significant underinvestment in energy efficiency. Other market failures, such as imperfect information and bounded rationality, almost certainly contribute to or exacerbate the uncertainty/loss aversion market deficiency. External costs, especially carbon dioxide emissions and oil dependence, are critically important market failures because they constitute society's motivation for seeking solutions to the fuel economy market failure.

References

American Council for an Energy Efficiency Economy (ACEEE). 2007. "Quantifying the Effects of Market Failures in the End-Use of Energy," Final draft report prepared for the International Energy Agency, by ACEEE, Washington, D.C., February.

An, F., D. Gordon, H. He, D. Kodjak and D. Rutherford. 2007. "Passenger Vehicle Greenhouse Gas and Fuel Economy Standards: A Global Update," International Council on Clean Transportation, Washington, D.C., July.

Benatzi, S. and R.H. Thaler. 1995. "Myopic Loss Aversion and the Equity Premium Puzzle," The Quarterly Journal of Economics, vol. 110, no. 1, pp. 73–92.

Congressional Budget Office (CBO). 2003. "The Economic Costs of Fuel Economy Standards Versus a Gasoline Tax," Congress of the United States, Washington, D.C., December.

Davis, S.C. and S.W. Diegel. 2007. Transportation Energy Data Book: Ed. 26, ORNL-6978, Oak Ridge National Laboratory, Oak Ridge, Tennessee.

Espey, M. 2005. "Do Consumers Value Fuel Economy?" Regulation, Winter 2005–2006, pp. 8–10. The conclusions in the above article are based on Espey, M. and S. Nair, 2005. "Automobile Fuel Economy: What is it Worth?", Contemporary Economic Policy, vol. 23, no. 3, pp. 317–323.

Fischer, C., W. Harrington and I.W.H. Parry. 2004. "Economic Impacts of Tightening the Corporate Average Fuel Economy (CAFE) Standards," report prepared for the Environmental Protection Agency and the National Highway Traffic Safety Administration by Resources for the Future, Washington, D.C., June 4, 2004.

Gal, D. 2006. "A Psychological Law of Inertia and the Illusion of Loss Aversion," Judgment and Decision Making, vol. 1, no. 1, pp. 23–32.

Goldstein, D.B. 2007. Saving Energy Growing Jobs, Bay Tree Publishing, Berkeley, California.

Greene, D.L., R. Goeltz, J. Hopson and E. Tworek. 2006. "Analysis of In-Use Fuel Economy Shortfall by Means of Voluntarily Reported Fuel Economy Estimates," Transportation Research Record No. 1983, Transportation Research Board, Washington, D.C.

Heavenrich, R.M. 2006. "Light-Duty Automotive Technology and Fuel Economy Trends: 1975 through 2006," EPA420-R-06-011, Office of Transportation and Air Quality, U.S. Environmental Protection Agency, Ann Arbor, Michigan, July.

Hopson, J.L. 2007. Personal communication, National Transportation Research Center, Knoxville, Tennessee, August 13.

Howarth, R.B. and A.H. Sanstad. 1995. "Discount Rates and Energy Efficiency," Contemporary Economic Policy, vol. 13, no. 3, pp. 101–109.

National Research Council (NRC). 2002. Effectiveness and Impact of Corporate Average Fuel Economy (CAFE) Standards, National Academies Press, Washington, D.C.

International Energy Agency (IEA). 2001. Saving Oil and Reducing CO2 Emissions in Transport, OECD, Paris.

International Energy Agency (IEA). 2000. "Policies and Measures to Mitigate GHG Emissions: Transportation Options (LDV) Technical Appendix," OECD, Paris, March.

International Energy Agency (IEA). 1991. Fuel Efficiency of Passenger Cars, OECD, Paris.

International Energy Agency (IEA) 1984. Fuel Efficiency of Passenger Cars, OECD, Paris.

Nye, R. 2002. "Qualitative Research Report: ORNL/NTRC Focus Groups," conducted by the Looking Glass Group, Knoxville, Tennessee, April 8, 2002.

Opinion Research Corporation. 2004. "CARAVAN ORC Study #713218 for the National Renewable Energy Laboratory," Princeton, New Jersey, May 20, 2004.

Opinion Research Corporation. 2007. "CARAVAN ORC Study 716159 for the National Renewable Energy Laboratory," Princeton, New Jersey, April 13, 2007.

Pizer, W. 2006. "The Economics of Improving Fuel Economy," Resources, Fall, Resources for the Future, Washington, D.C.

Parry, I.W.H., M. Walls and W. Harrington. 2007. "Automobile Externalities and Policies," Journal of Economic Literature, vol. XLV, pp. 373–399.

Rabin, M. 2000. "Diminishing Marginal Utility of Wealth Cannot Explain Risk Aversion," Institute of Business and Economic Research, Department of Economics, U.S. Berkeley, paper E00-287.

Rubenstein, A. 1998. Modeling Bounded Rationality, MIT Press, Cambridge, Massachusetts.

Sanstad, A.H. and R.B. Howarth. 1994. "Consumer Rationality and Energy Efficiency," Proceedings of the ACEEE 1994 Summer Study on Energy Efficiency in Buildings, Volume 1: Human Dimensions, American Council for an Energy Efficiency Economy, Washington, D.C.

Thaler, R.H., A. Tversky, D. Kahneman and A. Schwartz. 1997. "The Effect of Myopia and Loss Aversion on Risk Taking: An Experimental Test," The Quarterly Journal of Economics, vol. 112, no. 2, pp. 647–661.

Turrentine, T. and K. Kurani. 2005. "Automotive Fuel Economy in the Purchase and Use Decisions of Households," presented at the 84th Annual Meeting of the Transportation Research Board, National Research Council, January 9–13, Washington, D.C.

Tversky, A. and D. Kahneman. 1992. "Advances in Prospect Theory: Cumulative Representation of Uncertainty," Journal of Risk and Uncertainty, vol. 5, pp. 297–323.

U.S. Department of Energy, Energy Information Administration (DOE/EIA). 2007. International Energy Annual 2005, DOE/EIA-0219(2005), Washington, D.C., available at http://www.eia.doe.gov/iea .

U.S. Department of Energy, Energy Information Administration (DOE/EIA). 1996. Issues in Midterm Analysis and Forecasting 1996, DOE/EIA-0607(96), Washington, D.C.

Appendix A: Biographies of Editors and Authors

Editors

Daniel Sperling is Professor of Civil Engineering and Environmental Science and Policy, and founding Director of the Institute of Transportation Studies at the University of California, Davis (ITS-Davis). In February 2007, the governor of California appointed Dr. Sperling to the Board of the California Air Resources Board. He also served as co-director of the California Low Carbon Fuel Standard study, requested in the Governor's January 2007 Executive Order.

Dr. Sperling is recognized as a leading international expert on transportation technology assessment, energy and environmental aspects of transportation, and transportation policy. He was recently honored as a lifetime National Associate of the National Academies, is author or editor of 200 technical articles and 10 books, and has testified ten times to the US Congress and California Legislature on alternative fuels and advanced vehicle technology. He is co-chair of the Transportation Sustainability Committee of the Transportation Research Board of the National Academy of Science and advises senior executives of many automotive and energy companies, environmental groups, national governments, and DOE national laboratories.

He earned his Ph.D. in Transportation Engineering from the University of California, Berkeley (with minors in Economics and Energy & Resources) and his B.S. in Environmental Engineering and Urban Planning from Cornell University. Professor Sperling worked two years as an environmental planner for the U.S. Environmental Protection Agency and two years as an urban planner in the Peace Corps in Honduras.

James S. Cannon is an internationally recognized researcher specializing in energy development, environmental protection, and related public policy issues. He is President of Energy Futures, Inc., which he founded in 1979. Among its activities, Energy Futures publishes the quarterly international journal *The Clean Fuels and Electric Vehicles Report* and the bimonthly newsletter, *Hybrid Vehicles*. Mr. Cannon has written several books on alternative transportation fuels and advanced technologies, including **The Drive for Clean Air** and **Harnessing Hydrogen: The Key to Sustainable Transportation**. He has also edited two books, **The Hydrogen Transition,** and **Driving Climate Change: Cutting Carbon from Transportation**, based on previous Asilomar conferences.

Over the past two decades, Mr. Cannon's research into alternative transportation fuels has taken him to over 20 countries on 5 continents. He holds an AB degree in chemistry from Princeton University and an MS degree in biochemistry from the University of Pennsylvania. He lives with his family in Boulder, Colorado.

Authors

Anup Bandivadekar is a postdoctoral fellow in the Sloan Automotive Laboratory at the Massachusetts Institute of technology (MIT). His research interests are focused around developing frameworks and methods to foster innovative solutions towards achieving a more sustainable energy and transportation system. Currently, Anup is working with the MIT Energy Initiative on evaluating vehicle and fuel technologies that could significantly reduce greenhouse gas emissions and petroleum use from the U.S. light duty fleet over the next thirty years. He holds a Bachelor of Engineering degree from University of Mumbai and a Master of Science Degree from Michigan Technological University in the field of Mechanical Engineering as well as a Master of Science in Technology and Policy and a Ph.D. in Engineering Systems from MIT.

Rex Burkholder has spent the last two decades working as a teacher, community activist, and elected official to make the Portland area a more livable and environmentally healthy place. He was a founding member of the Bicycle Transportation Alliance, a non-profit group devoted to promoting cycling and improving cycling conditions, and the Coalition for a Livable Future. In 2000, he was elected to the Metro Council, where chairs the Joint Policy Advisory Committee on Transportation and focuses on policies to conserve energy, improve access and safety, and protect natural resources while boosting the region's economy.

David G. Burwell is a founding partner of the BBG Group, a transportation consulting firm specializing in sustainable transportation solutions. He has authored several papers on transportation, sustainability and land use. He served as Chair of the Transportation Research Board (TRB) Committee on Transportation and Sustainability, on the TRB panel the produced *Towards a Sustainable Future: Addressing the Long-Term Effects of Motor Vehicle Transportation on Climate and Ecology*, and on the TRB Executive Committee. He is a former President and CEO of the Surface Transportation Policy

Project (STPP) and the founding President and CEO of the Rails-to-Trails Conservancy (RTC). Mr. Burwell has a degree in Government from Dartmouth College, and a J.D. degree from the University of Virginia. He resides in Bethesda Maryland.

Lynette Cheah is a researcher at the Sloan Automotive Laboratory at the Massachusetts Institute of Technology (MIT), where she is currently pursuing a doctorate degree in Engineering Systems. Her research interest is in the life-cycle energy and environmental impacts, and material flow assessments in transportation systems. Her current project evaluates the feasibility and impact of passenger vehicle weight and size reductions in the U.S. Prior to returning to graduate school, she worked in the national agency for science and technology research in Singapore, overseeing the environmental science and engineering portfolio. Lynette holds a B.S. civil and environmental engineering from North-western University, and a M.S. in management science from Stanford University.

Gustavo Collantes is a Research Fellow in the Energy Technology Innovation Policy group at Harvard's Kennedy School of Government. His work focuses on energy and climate change policy, particularly as they relate to transportation, as well as the economics and politics of low-carbon energy alternatives. At the international level, he is working on the evaluation of climate impacts of the production of biofuels in developing countries, particularly in South America. Gustavo earned his Ph.D. in Transportation Technology and Policy from the University of California at Davis.

Philippe Crist is an Administrator at the Joint Transport Research Centre (JTRC) of the International Transport Forum (formerly the European Conference of Ministers of Transport) and the Organisation for Economic Co-operation and Development. He manages several international research projects bringing together government and academic experts from the 51 countries that comprise the International Transport Forum. Most recently, he co-ordinated a side-event on vehicle fuel efficiency at the United Nations Climate Change Conference in Bali, Indonesia (December 2007). He joined the JTRC in 2004. Prior to that he worked at the Maritime Transport Committee (from 2000) and Environment Directorate (from 1996) of the OECD. Mr. Crist was educated at Vanderbilt University and the French National Agronomic Institute.

Mark A. Delucchi is a research scientist at the Institute of Transportation Studies, University of California, Davis (ITS-Davis). His research includes comprehensive analyses of the full social-costs of motor-vehicle use, lifecycle analyses of emissions of greenhouse gases and criteria pollutants, modeling of the energy use and lifetime costs of advanced electric and conventional vehicles, and planning sustainable transportation systems for new communities.

Christopher Evans is a Masters of Science candidate in the Technology and Policy Program at the Massachusetts Institute of Technology. He works in the Sloan Automotive Laboratory, investigating how policy options and new technologies can contribute to reductions in greenhouse gas emissions and fuel use in the U.S. light duty vehicle fleet. Christopher received his Bachelors of Science in Mechanical Engineering from the University of Manitoba.

Carolyn Fischer is a Senior Fellow at Resources for the Future (RFF), an independent environmental policy research institute based in Washington, DC. Her research addresses policy mechanisms and modeling tools that cut across environmental issues, including environmental policy design and technological change, international trade and environmental policies, and resource economics. Her climate policy work has focused on the role of allocation mechanisms in emissions trading design and the opportunities and challenges posed by international trade. Fischer holds a Ph.D. in Economics from the University of Michigan and a B.A. in International Relations from the University of Pennsylvania. Previously, she has taught at Johns Hopkins University and served as a staff economist for the Council of Economic Advisors to the President. She currently sits on the Board of Directors of the Association of Environmental and Resource Economists.

Kelly Sims Gallagher, Adjunct Lecturer in Public Policy, is Director of the Energy Technology Innovation Policy (ETIP) research group at the Harvard Kennedy School's Belfer Center for Science and International Affairs. Her work is focused on studying, informing, and shaping U.S. and Chinese energy and climate-change policy. U.S.-China energy cooperation and energy technology innovation, including technology transfer, are important themes in her work. She has specialized particularly on energy policy related to transportation and coal in both countries. She is currently serving on the Task Force on Innovation for the China Council for International Cooperation on Environment and Development, and as Counselor-at-Large for the Asia Society-Council on Foreign Relations-Brookings Institution Initiative for U.S.-China Cooperation on Energy and Climate. She previously worked for Fluor Daniel Environmental Services, the Office of Vice President Al Gore, and Ozone Action. A Truman Scholar, she has a MALD and PhD in international affairs from the Fletcher School at Tufts University, and an AB in international relations and environmental studies from Occidental College.

John German is Manager of Environmental and Energy Analyses for American Honda Motor Company. His responsibilities include anything connected with environmental and energy matters, with an emphasis on being a liaison between Honda's R&D people and regulatory affairs. Mr. German has been involved with advanced technology and fuel economy since joining Chrysler in 1976, where he spent 8 years in Powertrain Engineering working on fuel economy issues. Prior to joining Honda 10 years ago, he spent 13 years doing research and writing regulations for EPA's Office of Mobile Sources' laboratory in Ann Arbor, MI. Mr. German is the author of a variety of technical papers and a book on hybrid electric vehicles published by SAE. He was the first recipient of the Barry D. McNutt award, presented annually by SAE for Excellence in Automotive Policy Analysis. He has a bachelor's degree in Physics from the University of Michigan and got over half way through an MBA before he came to his senses.

David L. Greene is a Corporate Fellow in Oak Ridge National Laboratory's Energy and Transportation Science Division, David Greene has spent 30 years researching transportation energy and environmental policy issues for the federal government. Dr. Greene has authored or co-authored more than two hundred professional publications, including over seventy-five articles in refereed journals. He was the first Editor-in-Chief of the *Journal of Transportation and Statistics* and serves on the editorial boards of *Energy Policy* and *Transportation Research D*. He was a lead author for the Second, Third and Fourth Assessment Reports of the Intergovernmental Panel on Climate Change and a lead author for the IPCC report on Aviation and the Global Atmosphere. Dr. Greene earned a B.A. degree from Columbia University in 1971, an M.A. from the University of Oregon in 1973, and a Ph.D. in Geography and Environmental Engineering from The Johns Hopkins University in 1978. In recognition of his service to the National Research Council, Dr. Greene was designated a lifetime National Associate of the U.S. National Academies. A 20-year member of the Society of Automotive Engineers, David Greene was the recipient of the SAE's 2007 Barry D. McNutt Award for Excellence in Automotive Policy Analysis.

Anthony Greszler is Vice President Advanced Engineering, Volvo Powertrain North America. He has been involved with diesel engine design and development since 1977. His diesel experience includes all mechanical systems, cooling, lubrication, performance development, emissions, controls, and advanced concepts. He has also been involved with heavy duty natural gas engines and other fuel alternatives, particularly DME. From 1977-2001 he was with Cummins Engine Co. responsible for design and development of heavy duty diesel engines, including 2 years in Europe on N14 and L10 engines and 8 years as L10 & M11 Chief Engineer, including on-highway and off-highway applications. In 2001, he became Vice President, Engineering for Volvo Powertrain, North America with responsibility for engine development for Mack Trucks and Volvo Trucks North America, including Mack ETECH, ASET, and E7 natural gas engine, support for Volvo D12 in North America, and development for future North American engines including US 2007 and 2010 emissions. In 2005, he took responsibility for Advanced Engineering for Engines and Vehicle Propulsion with focus on diesel combustion/emissions, hybrid propulsion, and alternative fuels. He serves on the Executive Council of the Engine Manufacturers Association.

John Heywood has been a faculty member at MIT since 1968, where he is Director of the Sloan Automotive Laboratory and Sun Jae Professor of Mechanical Engineering. His research is focused on internal combustion engines, their fuel requirements, and broader studies of future transportation technology, fuels, and emissions. He has published extensively in the technical literature, holds a number of patents, and is the author of a major text and professional reference "Internal Combustion Engine Fundamentals." He is a Fellow of the Society of Automotive Engineers. He received a 1996 U.S. Department of Transportation Award for the Advancement of Motor Vehicle Research and Development. He is a member of the National Academy of Engineering and a Fellow of the American Academy of Arts and Sciences. He has honorary degrees from Chalmers University of Technology, Sweden, and City University, London.

Amy Myers Jaffe, a Princeton University graduate in Arabic Studies, is the Wallace S. Wilson fellow for Energy Studies at the James A. Baker III Institute for Public Policy and associate director of the Rice University energy program. Ms. Jaffe's research focuses on the subject of oil geopolitics, strategic energy policy, and energy economics. Ms. Jaffe is widely published in academic journals and numerous book volumes including: a co-authored keynote article in the *The Whitehead Journal of Diplomacy and International Relations*, "Energy Security: Meeting the Growing Challenge of National Oil Companies," (Summer 2007); and *Survival*, "The Persian Gulf and the Geopolitics of Oil," (Spring 2006). She served as co-editor of two books: *The Geopolitics of Natural Gas* (Cambridge University Press, 2005) and *Energy in the Caspian Region: Present and Future* (Palgrave, 2002). She has been named to Who's Who in America, 2008 and was among Esquire Magazine's 100 Best and Brightest honorees in 2005. Jaffe served as expert on the Baker/Hamilton Iraq Study Group and served as project director for the Baker Institute/Council on Foreign Relations task force on Strategic Energy Policy. She is currently serving as a strategic advisor to the American Automobile Association (AAA) of the United States helping the motor club fashion a voice for the American motorist in the U.S. energy policy debate. Prior to joining the Baker Institute, Ms. Jaffe was the senior editor and Middle East analyst for Petroleum Intelligence Weekly, a respected oil journal.

Nic Lutsey just completed his Ph.D. in the Transportation Technology and Policy program at the Institute of Transportation Studies at the University of California, Davis (ITS-Davis). He holds a Bachelor of Science in engineering from Cornell University. Over his time at ITS-Davis, his work has included the assessment of technologies and policies to promote cost-effective energy efficiency and emission reduction opportunities in passenger and commercial vehicles. He has co-authored numerous reports and nine peer-reviewed articles on energy and climate change topics.

Eliot Rose began working as a policy associate at Metro, the Portland, Oregon, U.S.A. area's regional government, in 2007. Prior to that, he taught middle school. While teaching a sustainability unit to 7th graders he realized that many adults also needed education on the topic. Eliot began to volunteer for the Coalition for a Livable Future, which unites over 90 organizations with the common goal of creating a more sustainable region, speaking to community groups about the hidden costs and benefits of transportation choices. He continues to explore this topic, as well as renewable energy options and climate change mitigation, in his work at Metro.

Jack Short is the Secretary General of the International Transport Forum. The International Transport Forum was set up by Transport Ministers of 51 countries in 2006 to foster global strategic discussions on Transport for Ministers, key industry and societal actors. The International Transport Forum is the 'successor' to the ECMT which acted for many years as a "Think Tank" for Transport Ministries. It has a Secretariat of 25 people and is administratively part of the OECD in Paris. He is also the Director of the Joint OECD/ITF Transport Research Center, which was set up in 2004. He joined the ECMT in 1984 and was Deputy Secretary General from 1993 to 2001. Previously he worked for the Ministries of Transport and Finance in Ireland and in Transport Research. Mr. Short was educated at University College, Cork, and Trinity College, Dublin where he obtained masters degrees in Mathematics and Statistics.

Kurt Van Dender is an Administrator at the Joint Transport Research Centre (JTRC) of the International Transport Forum and the Organization for Economic Co-operation and Development, and an Associate Professor of Economics at the University of California, Irvine. He analyses and follows up on policies to mitigate transport externalities and on issues relating to the governance of congestion-prone transport facilities. His work is published in leading transport and economics journals. He holds a Ph.D. from the Catholic University of Leuven, Belgium.

Appendix B: Asilomar 2007 Attendee List

Hayato Akizuki	Nissan
Fabian Allard	Natural Resources Canada
Feng An	Ameritech
Don Anair	Union of Concerned Scientists
Anup Bandivadekar	MIT
Nicole Barber	Chevron
William Barron	Hong Kong University of Science & Technology
Louise Bedsworth	Public Policy Istitute of California
Anthony Bernhardt	Environmental Entrepreneurs
Robert Bienenfeld	Honda
Steven Bimson	Center for Sustainable Energy California
Jessica Bird	California Legislative Analyst's Office
KC Bishop III	Chevron
William Black	Indiana University
Carl Blumstein	UC Office of the President
Raymond Boeman	Oak Ridge National Laboratory
John Boesel	CALSTART
Andre Bourbeau	Transport Canada
Bill Boyce	Sacramento Municipal Utility District
James Boyd	California Energy Commission
Joe Browder	Dunlap & Browder, Inc.
Susan Brown	California Energy Commission
David Brownstone	UC Irvine
David Bunch	UC Davis
Andrew Burke	UC Davis
Rex Burkholder	Portland Metro Council
David Burwell	BBG Group
Joshua Bushinsky	The Pew Center
Jeffrey Byron	California Energy Commission
John Cabaniss	Association of International Automobile Manufacturers
Tom Cackette	California Air Resources Board

Jim Cannon	Energy Future, Inc.
Tim Carmichael	Coalition for Clean Air
Edie Chang	California Air Resources Board
Elaine Chang	South Coast Air Quality Management District
William Chernicoff	U.S. DOT
Joy Chiu	New York State DOT
Sue Cischke	Ford Motor Company
Michael Coates	Mighty Communications
Douglas Comeau	Valero Energy Corporation
James Corbett	University of Delaware
Cynthia Cory	California Farm Bureau Federation
William Cowart	Cambridge Systematics
William Craven	DaimlerChrysler
Greg Dana	Auto Alliance
John DeCicco	Environmental Defense
Mark Delucchi	UC Davis
Harald Diaz-Bone	UN Climate Change Secretariat
Jay Dickenson	California Legislative Analyst's Office
Clarence Ditlow	Center for Auto Safety
Robert Dixon	International Energy Agency
Bill Drumheller	Oregon DOE
Kelly Dunlap	Caltrans
Louise Dunlap	Dunlap & Browder, Inc.
Catherine Dunwoody	California Fuel Cell Partnership
George Eads	CRA International
Michael Eaves	California Natural Gas Vehicle Coalition
Jill Egbert	PG&E
Duncan Eggar	BP
Anthony Eggert	California Air Resources Board
Bob Epstein	Enivronmental Entrepreneurs
Mark Evers	Transport For London
Yueyue Fan	UC Davis
Alex Farrel	UC Berkeley
Malcolm Fergusson	Institute for European Environmental Policy
Charles Fielder	Caltrans
Carolyn Fischer	Resources for the Future
William Fitzharris	BP
Jesse Fleming	Natural Resources Canada
Scott Folwarkow	Valero Energy Corporation
Emil Frankel	Bipartisan Policy Center
Mary Frederick	Caltrans
Sally French	California Integrated Waste Management Board
Danielle Fugere	Friends of the Earth
Tom Fulks	Mighty Communications
Shuk Wai Freda Fung	Environmental Defense

Cynthia Gage	U.S. EPA
Julia Gamas	U.S. EPA
John German	Honda
John Ginder	Ford
Garry Gordan	Sacramento Area Council of Governments
Kevin Green	The Volpe National Transportation Systems Center
Ellen Greenberg	UC Davis
David Greene	Oak Ridge National Laboratory
Larry Greene	Sacramento Metropolitan Air Quality Management District
Anthony Greszler	Volvo
Charles Griffith	Ecology Center of Ann Arbor
Chris Grundler	U.S. EPA
Taro Hagiwara	Nissan
Wilhelm Hall	BMW
Donald Hardesty	Sandia National Laboratory
Brenda Hensler-Hobbs	Transport Canada
John Heywood	MIT
Ed Hillsman	Washington State DOT
Toshio Hirota	Nissan
Ananda Hirsch	Energy Foundation
Yoshiaki Hitomi	Nissan
John Horsley	AASHTO
Jamie Hulan	Transport Canada
John Hutchison	Ontario Ministry of Environment
Roland Hwang	Natural Resources Defense Council
Rahul Iyer	Primafuel, Inc.
Michael Jackson	TIAX LLC - Acurex
Jeffrey Jacobs	Chevron
Norman Johnson	Robert Bosch Corporation
Jack Johnston	ExxonMobil (retired)
Brian Johnston	Nissan
Hal Kassoff	Parsons Brincherhoff
Dean Kato	Toyota
Jay Keller	Sandia National Laboratories
Alissa Kendall	UC Davis
Paul Khanna	Natural Resources Canada
Jamie Knapp	Jamie Knapp Communications
Ben Knight	Honda
Bob Knight	Bevilacqua-Knight, Inc.
Christopher Knittel	UC Davis
Reinhart Kuehne	DLR - Verkehrsstudien
Stephen Kukucha	Ballard Power Systems, Inc.
Ken Kurani	UC Davis
Robert Larsen	Argonne National Laboratory

Bob Larson	U.S. EPA
Michael Lawrence	Jack Faucett Associates
Sungwon Lee	The Korean Transport Institute
Martin Lee-Gosselin	Universite Laval
Paul Leiby	Oak Ridge National Lab
Zheng Li	Tsinghua-BP Energy Centre
C.Y. Cynthia Lin	UC Davis
Timothy Lipman	UC Berkeley
Chung Liu	South Coast Air Quality Management District
Hengwei Liu	Tsinghua University
Jane Long	Lawrence Livermore National Laboratory
Michael Lord	Toyota
Deron Lovaas	Natural Resources Defense Council
Michael Love	Toyota
Amy Luers	Union of Concerned Scientists
Jason Mark	Energy Foundation
David Marler	ExxonMobil
Scott Mason	ConocoPhillips
Alan McKinnon	Heriot - Watt University
Alan Meier	UC Davis and LBNL
Thomas Menzies	Transportation Research Board
Russell Meyer	ICF International
Martine Micozzi	Transportation Research Board
Ron Milam	Fehr & Peers
Shannon Miles	Natural Resources Canada
Marianne Mintz	Argonne National Laboratory
Philip Misemer	California Energy Commission
Patricia Monahan	Union of Concerned Scientists
Ralph Moran	BP
Amy Myers Jaffe	Baker Institute for Public Policy
Reza Navai	Caltrans
Mary Nichols	California Air Resources Board
Paul Nieuwenhuis	Cardiff University
Joan Ogden	UC Davis
Victoria Orsborne	Natural Resources Canada
Munehiko Oshima	Nissan
Neil Otto	Ballard Automotive (retired)
George Parks	ConocoPhillips
Mark Paster	U.S. DOE
Richard Plevin	UC Berkeley
Steven Plotkin	Argonne National Laboratory
Joel Pointon	Sempra Energy
Joanne Potter	Cambridge Systematics
Jim Presswood	Natural Resources Defense Council
Jorge Prozzi	University of Texas

Jim Ragland	Aramco Services Company
David Raney	Honda
Catherine Reheis-Boyd	Western States Petroleum Association
Peter Reilly-Roe	Marbek Resources Consultants
Michael Replogle	Environmental Defense
Peter Rohde	EnergyWashington
Jack Rosebro	Perfect Sky
Jonathan Rubin	University of Maine
Barney Rush	H2Gen Innovations, Inc.
Ichiro Sakai	Honda
Amul Sathe	Natural Resources Defense Council
Michael Savonis	U.S. DOT
Robert F. Sawyer	UC Berkeley
Lee Schipper	World Resources Institute
Susan Schoenung	Longitude 122 West, Inc.
Marcy Schwartz	CH2M Hill
Mark Schwartz	PIRA Energy
Peter Schwartz	Global Business Network
Paul Scott	ISE Corporation
Susan Shaheen	UC Berkeley
Rosella Shapiro	California Energy Commission
John Shears	Center for Energy Efficiency & Renewable Technology
Jack Short	International Transport Forum
Harry Sigworth	Chevron
Fred Silver	CALSTART
Dean Simeroth	California Air Resources Board
Steven Skerlos	University of Michigan
Chris Sloane	GM
Gail Slocum	PG&E
Robert Sorrell	BP
Quong Spencer	Union of Concerned Scientists
Dan Sperling	UC Davis
Wolfgang Steiger, Dr.	Volkswagen
Irene Stillings	San Diego Environmental Foundation
Laura Stuchinsky	Silicon Valley Leadership Group
Dan Sturgis	University of Colorado at Boulder
Jane Summerton	VTI, Swedish National Road & Transport Research Institute
George Sverdrup	National Renewable Energy Laboratory
Graeme Sweeney	Shell
Fujio Takimoto	Subaru
Ruth Talbot	Natural Resources Canada
Lindsee Tanimoto	Caltrans
Margaret Taylor	UC Berkeley

Laurie ten Hope	California Energy Commission
Sven Thesen	PG&E
John Tillman	Volkswagen
Luke Tonachel	Natural Resources Defense Council
John Topping	Climate Institute
Gary Toth	The Project for Public Spaces
Jan Tribulowski	BMW
Andreas Truckenbrodt	DaimlerChrysler
Brian Turner	International Council on Clean Transportation
Tom Turrentine	UC Davis
Stefan Unnasch	Life Cycle Associates
Peter Ward	California Energy Commission
Mia Waters	Washington State DOT
Hank Wedaa	California Hydrogen Business Council
Thomas White	U.S. DOE
Jill Whynot	South Coast Air Quality Management District
Stephanie Williams	Western Energy Institute
Jon Williams	Transportation Research Board
John Wilson	California Energy Commission
James Winebrake	Rochester Institute of Technology
Steve Winkelman	Center for Clean Air Policy
Robert Wooley	Abengoa Bioenergy
Brian Wynne	Electric Drive Transportation Association
Christopher Yang	UC Davis
Jo-Ann Yantzis	Clean Energy Fuel
Phyllis Yoshida	U.S. DOE
Rick Zalesky	Chevron
John Zamurs	New York State DOT
Bill Zobel	Sempra Energy
Jeffrey Zupan	Transportation for Regional Plan Association

Index